Undergraduate Texts in Mathematics

Springer
New York
Berlin
Heidelberg
Barcelona
Hong Kong
London
Milan
Paris
Singapore
Tokyo

Undergraduate Texts in Mathematics

Abbott: Understanding Analysis.
Anglin: Mathematics: A Concise History and Philosophy.
Readings in Mathematics.
Anglin/Lambek: The Heritage of Thales.
Readings in Mathematics.
Apostol: Introduction to Analytic Number Theory. Second edition.
Armstrong: Basic Topology.
Armstrong: Groups and Symmetry.
Axler: Linear Algebra Done Right. Second edition.
Beardon: Limits: A New Approach to Real Analysis.
Bak/Newman: Complex Analysis. Second edition.
Banchoff/Wermer: Linear Algebra Through Geometry. Second edition.
Berberian: A First Course in Real Analysis.
Bix: Conics and Cubics: A Concrete Introduction to Algebraic Curves.
Brémaud: An Introduction to Probabilistic Modeling.
Bressoud: Factorization and Primality Testing.
Bressoud: Second Year Calculus.
Readings in Mathematics.
Brickman: Mathematical Introduction to Linear Programming and Game Theory.
Browder: Mathematical Analysis: An Introduction.
Buchmann: Introduction to Cryptography.
Buskes/van Rooij: Topological Spaces: From Distance to Neighborhood.
Callahan: The Geometry of Spacetime: An Introduction to Special and General Relavitity.
Carter/van Brunt: The Lebesgue–Stieltjes Integral: A Practical Introduction.
Cederberg: A Course in Modern Geometries. Second edition.
Childs: A Concrete Introduction to Higher Algebra. Second edition.
Chung: Elementary Probability Theory with Stochastic Processes. Third edition.
Cox/Little/O'Shea: Ideals, Varieties, and Algorithms. Second edition.
Croom: Basic Concepts of Algebraic Topology.
Curtis: Linear Algebra: An Introductory Approach. Fourth edition.
Devlin: The Joy of Sets: Fundamentals of Contemporary Set Theory. Second edition.
Dixmier: General Topology.
Driver: Why Math?
Ebbinghaus/Flum/Thomas: Mathematical Logic. Second edition.
Edgar: Measure, Topology, and Fractal Geometry.
Elaydi: An Introduction to Difference Equations. Second edition.
Exner: An Accompaniment to Higher Mathematics.
Exner: Inside Calculus.
Fine/Rosenberger: The Fundamental Theory of Algebra.
Fischer: Intermediate Real Analysis.
Flanigan/Kazdan: Calculus Two: Linear and Nonlinear Functions. Second edition.
Fleming: Functions of Several Variables. Second edition.
Foulds: Combinatorial Optimization for Undergraduates.
Foulds: Optimization Techniques: An Introduction.
Franklin: Methods of Mathematical Economics.
Frazier: An Introduction to Wavelets Through Linear Algebra.
Gamelin: Complex Analysis.
Gordon: Discrete Probability.
Hairer/Wanner: Analysis by Its History.
Readings in Mathematics.
Halmos: Finite-Dimensional Vector Spaces. Second edition.

(continued after index)

John Stillwell

Elements of Algebra

Geometry, Numbers, Equations

With 34 illustrations

Springer

John Stillwell
Department of Mathematics
Monash University
Clayton, Victoria 3168
Australia

Mathematics Subject Classifications (2000): 00-01, 08-01, 115XX, 14-XX

Library of Congress Cataloging-in-Publication Data
Stillwell, John.
Elements of algebra : geometry, numbers, equations / John Stillwell.
p. cm. – (Undergraduate texts in mathematics)
Includes bibliographical references and index.
ISBN 978-1-4419-2839-9
1. Algebra. I. Title. II. Series.
QA155.S75 1994
512'.02-dc20 94-10085

Printed on acid-free paper.

Production managed by Karen Phillips; manufacturing supervised by Vincent Scelta.
Photocomposed pages prepared from the author's LaTeX files.

Printed in the United States of America.

9 8 7 6 5 4 3 (Corrected third printing, 2001)

Springer-Verlag New York Berlin Heidelberg
A member of BertelsmannSpringer Science+Business Media GmbH

To Elaine

Preface

Algebra is abstract mathematics – let us make no bones about it – yet it is also applied mathematics in its best and purest form. It is not abstraction for its own sake, but abstraction for the sake of efficiency, power and insight. Algebra emerged from the struggle to solve concrete, physical problems in geometry, and succeeded after 2000 years of failure by other forms of mathematics. It did this by exposing the mathematical structure of geometry, and by providing the tools to analyse it. This is typical of the way algebra is applied; it is the best and purest form of application because it reveals the simplest and most universal mathematical structures.

The present book aims to foster a proper appreciation of algebra by showing abstraction at work on concrete problems, the classical problems of construction by straightedge and compass. These problems originated in the time of Euclid, when geometry and number theory were paramount, and were not solved until the 19th century, with the advent of abstract algebra. As we now know, algebra brings about a unification of geometry, number theory and indeed most branches of mathematics. This is not really surprising when one has a historical understanding of the subject, which I also hope to impart.

The bridge between Euclid and abstract algebra is the algebraic geometry invented by Fermat and Descartes around 1630. By assigning numerical coordinates to points in the plane, they were able to restate many geometric problems as problems about polynomial equations. Thanks to 16th century advances in the treatment of equations, they found it easy to solve many problems, some of which had defeated the ancients. However, certain problems remained intractable, particularly those involving equations of degree ≥ 3. At first it was thought that improved technique would solve these too, but as time went by this hope faded, and a fundamental shift in thinking took place. By the end of the 18th century, mathematicians were considering the possibility of equations *without* solutions, or at least without solutions of a certain geometric type ("constructible" solutions).

Such a possibility calls for a more abstract level of algebraic thought. Instead of treating equations, it is necessary to treat *properties* of equations, since the goal is to recognise the property of solvability. Since the equations are polynomial equations, this amounts to studying the properties of polynomials. This was done very successfully by Lagrange, Gauss, Abel and Galois between 1770 and 1830. Their work was successful not only in solving the ancient construction problems, but also in creating the concepts that are the backbone of algebra today – rings, fields, and groups.

However, these concepts were not very clear at the time. They were identified and disentangled from the old theory of equations only a century later, through the work of Jordan, Kronecker, Dedekind, Noether and Artin, and first presented to the general mathematical public in the *Moderne Algebra* of van der Waerden [1931]. The result was an algebra which, ironically, could live without geometry – and to some mathematicians that meant an algebra which *should* live without geometry. Other mathematicians (and students!) were alienated and bewildered by this development, particularly those who were geometrically inclined. It is true that algebra was separated from geometry with the best of intentions. Some of the leading "separatists," in fact, were geometers who wanted to build rigorous foundations for algebraic geometry and topology, and this called for an algebra without any geometric assumptions. All the same, separation of algebra from geometry was a pedagogical mistake, and fortunately one from which we are beginning to recover.

Since the great virtue of algebra is its power to unify topics from number theory to geometry, why not develop the subject with unification in mind? I hope that the present book demonstrates that such a unification is possible. It grew out of a course in Galois theory for 3rd year students at Monash University, but should in principle be accessible to anyone with a strong secondary school background, assuming this background includes the language of sets and functions. On the other hand, it should also be of interest to mathematicians who know the technicalities of abstract algebra but wish to know more about its historical context. Although it is no substitute for a comprehensive history of algebra (which has yet to be written), this book does try to locate the sources of the main ideas. They can be picked up on the fly from the references to the original literature, in the name [year] format, or mulled over in the end-of-chapter discussions. Since the aim of the book is to lead the reader to better things, I hope these discussions will open the door to the great works by Gauss, Abel, Galois and others. Looking to the future as well as the past, there is also some discussion of recent developments and open problems.

The book is divided into sections small enough to be digested in one sitting. There is at most one theorem per section, so theorems can be identified by their section numbers. For example, Theorem 8.4 refers to the theorem in Section 8.4. The most frequently used theorems have been given names rather than numbers, so that it will be easier to recall what they are about. Exercises are placed in small groups at the end of sections, in the hope that they will be more tempting, and less difficult, when the reader's mind is already on the right track. They include many interesting theorems that could not be squeezed into the main text. The starred sections contain material that can be omitted from a basic course. If time is short (as it is at Monash, where we have only 24 lectures), then Chapters 3 and 9 can also be omitted.

I would like to thank Emma Carberry, Angelo di Pasquale, Helena Gregory, Mark Kisin, Sean Lucy, Greg Pantelides, Karen Parshall, Abe Shenitzer, Tanya Staley and Drew Vandeth for many corrections and improvements, Anne-Marie Vandenberg for her usual splendid typing, and my wife Elaine for her sharp-eyed proofreading.

Clayton, Victoria, Australia John Stillwell

Contents

1 Algebra and Geometry

1.1 Algebraic Problems

The purpose of this chapter is to sketch the classical background of abstract algebra. I shall introduce some famous mathematical problems, translate them into problems about equations, then try to indicate the difficulties that stand in the way of their solution. One reason for doing this is to demonstrate that algebra is central to mathematics. It is the natural setting for some of the oldest mathematical problems and, as we shall eventually see, it provides the keys to their solution. A second reason, complementary to the first, is to demonstrate the continuing relevance of classical mathematics. Even ancient mathematics is not dead or obsolete – it stays alive because its problems continue to stimulate the creation of new concepts and techniques. The most fertile problems in mathematics are over 2000 years old and still have not yielded up all their secrets.

The ancient mathematics I am talking about is encapsulated in Euclid's *Elements*, written around 300 BC to systematise the mathematics then known. Euclid's systematisation was so masterly that the *Elements* became the basic text for almost all mathematicians born before the 20th century. The fact that it no longer plays such a role is due to the tremendous growth of mathematics in the 19th century – particularly the growth of algebra. The *Elements* contains no algebra as we know it, though it does contain results that can be viewed as algebra with hindsight. In fact, the algebraic viewpoint reveals a hidden unity between different parts of the *Elements* that Euclid himself could not have foreseen. Thus both algebra and Euclid stand to gain by pooling their resources, as I hope to show in the present book.

I shall begin by explaining what polynomials have to do with geometry. The first step is to explain the connection between geometric constructions and the fundamental operations of arithmetic.

1.2 Straightedge and Compass Constructions

The simplest geometric instruments are the *straightedge*, used to draw straight lines, and *compass*, used to draw circles. More precisely, the straightedge is used to draw the line through a given pair of points, and the compass is used to draw the circle with a given center (where the "point" is placed) and passing through another given point (where the pencil or pen tip is placed). Thus if certain points are initially given, further points may be *constructed* as the intersections of lines or circles through the initial points, then further lines and circles may be constructed through the newly constructed points, and so on.

The distances between the initially given points may be regarded as given numbers, for example by taking one of the distances arbitrarily as the unit of length. It then turns out to be possible to construct, as a distance between constructible points, any number obtainable from the given ones by the operations $+, -, \times, \div$ (by a nonzero number) and $\sqrt{}$ (of a positive number). Of course, a sign convention is needed in order to interpret certain distances as negative.

Given lengths a and b, it is clear how to construct $a+b$ and $a-b$. Namely, use the straightedge to construct a line containing the length a, then use the compass to transfer the length b to this line, adding and subtracting b by means of a circle of radius b about one end of a (Figure 1.2.1).

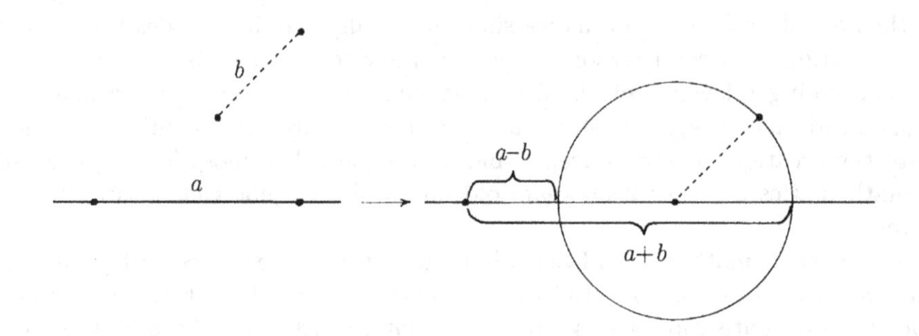

Fig. 1.2.1. Addition and subtraction of lengths

The construction of ab and a/b depends on the construction of a perpendicular M to a line L at a point P, which can be done as in Figure 1.2.2. It is then possible to draw the pairs of right-angled triangles shown in Figure 1.2.3, whose similarity shows that ab and a/b are the lengths indicated.

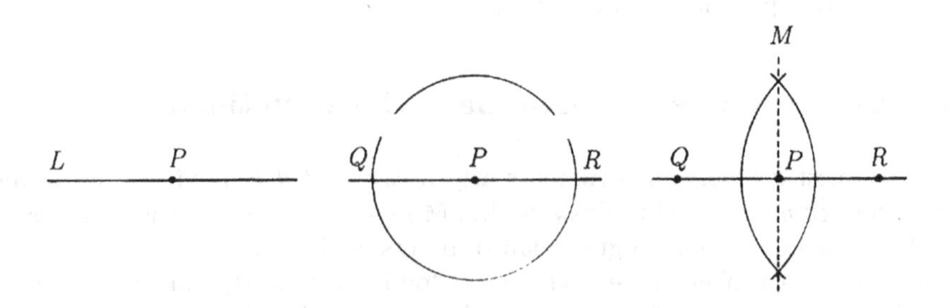

Fig. 1.2.2. Construction of a perpendicular

To construct $\sqrt{a}$ we draw Figure 1.2.4, using the process of Figure 1.2.2 again to bisect the length $a+1$ and hence draw the circle with diameter $a+1$. The altitude of $\sqrt{a}$ can be explained by the fact that all three triangles in Figure 1.2.4 are similar right-angled triangles, since the angle at A is also a right angle. (The latter fact is explained by Figure 1.2.5, where the equal radii yield isosceles triangles, hence angles like those shown, which imply $\alpha + \beta = \pi/2$.)

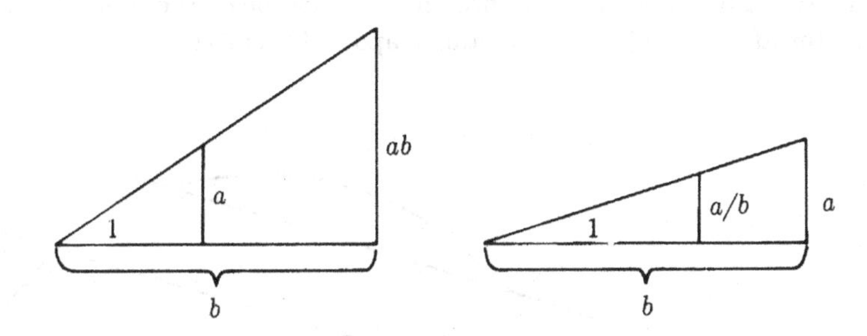

Fig. 1.2.3. Multiplication and division of lengths

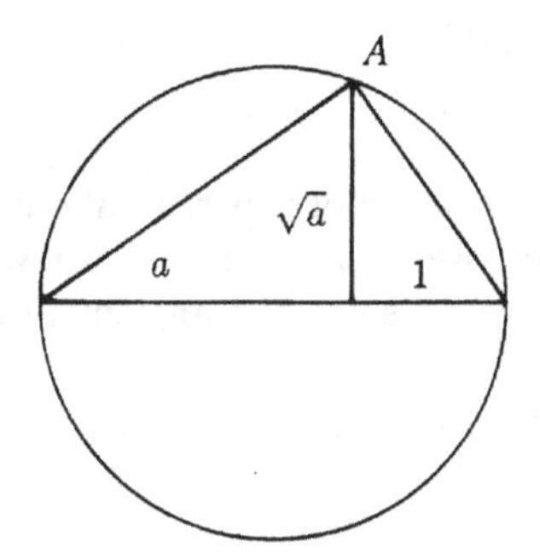

Fig. 1.2.4. Square root construction

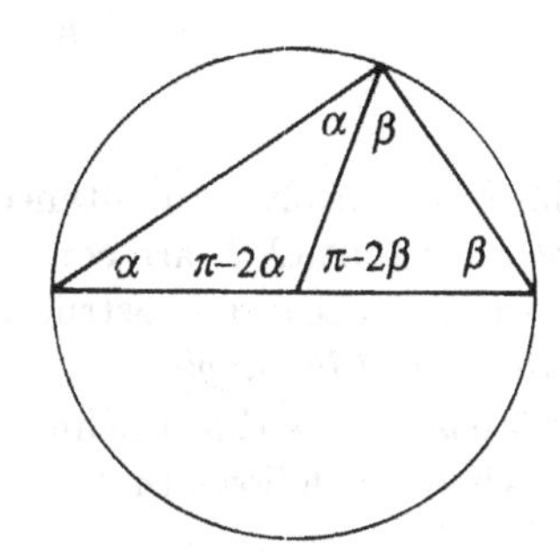

Fig. 1.2.5. Right angle in semicircle

Now the numbers obtainable from 1 by $+, -, \times, \div$ form the set of rational numbers $\mathbb{Q}$. We shall call the set of numbers obtainable from $\mathbb{Q}$ by square roots of positive members as well as $+, -, \times, \div$ the *real quadratic closure of* $\mathbb{Q}$. Thus the numbers $\sqrt{3}$, $2 - \sqrt{3}, \sqrt{2 - \sqrt{3}}$ all belong to the real quadratic closure of $\mathbb{Q}$.

Let us also call a number d *constructible* if there are points, constructible from a given pair unit distance apart, whose distance apart is d. Then the argument of this section shows that *each number in the real quadratic closure of* $\mathbb{Q}$ *is constructible.* In Section 1.3 we shall prove the converse of this proposition, thus obtaining an algebraic characterisation of the constructible numbers.

Remarks. The reason we say that the line drawing instrument is a straightedge, rather than a ruler, is to avoid any misunderstanding about the "construction" of lines. A straightedge has no marks on it, so all it can do is draw lines between previously constructed or given points, as specified in the definition of constructibility. A ruler, on the other hand, has marks which apparently enable us to find a line without knowing more than one point on it.

For example, if the ruler has marks unit distance apart, then it can be used as in Figure 1.2.6 to find the line through P which meets the unit semicircle and its diameter at points Q, R unit distance apart. Or can it?

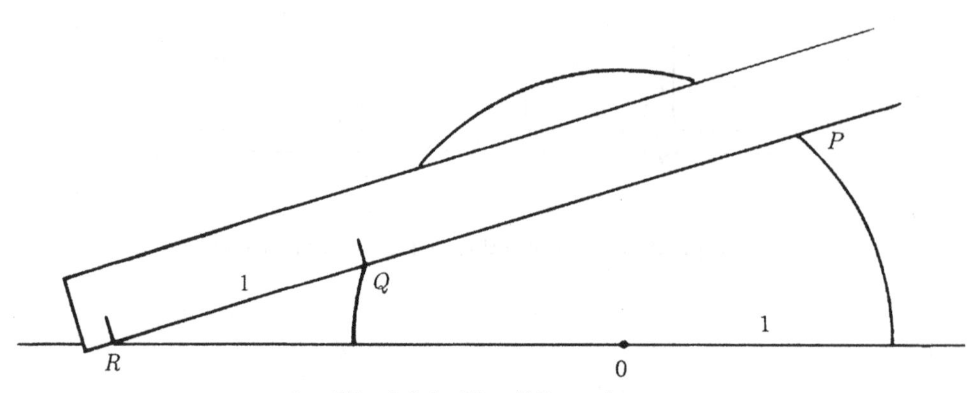

Fig. 1.2.6. The sliding ruler

Since this involves simultaneous sliding and rotation of the ruler about P, it is *not* clear that Q, R are constructible. In fact, we shall eventually show that they are not, because construction of Q or R would solve the ancient problem of *trisection of the angle.*

It is easy to see that the angles in Figure 1.2.6 are related as shown in Figure 1.2.7, whence it follows that

$$\beta = \pi - \gamma - \alpha = \pi - 4\gamma,$$

and hence $\gamma = \alpha/3$. This "trisection" of a given angle is attributed to Archimedes (284-212 BC). However, the classical trisection problem is to *construct* $\alpha/3$ for given α (Section 1.5), and we shall see that this is generally impossible (Section 5.5). It follows that the points Q, R are generally not constructible, and thus the line PQR "found" by ruler is not a constructible line.

Of course, one is free to study generalisations of the notion of constructibility which involve the use of rulers and other instruments. An absorbing and quite elementary account of this field may be found in Bieberbach [1952]. The fact remains, however, that the classical notion of constructibility has been the most fruitful for the development of algebra. The simple algebraic characterisation of constructible numbers has a lot to do with this.

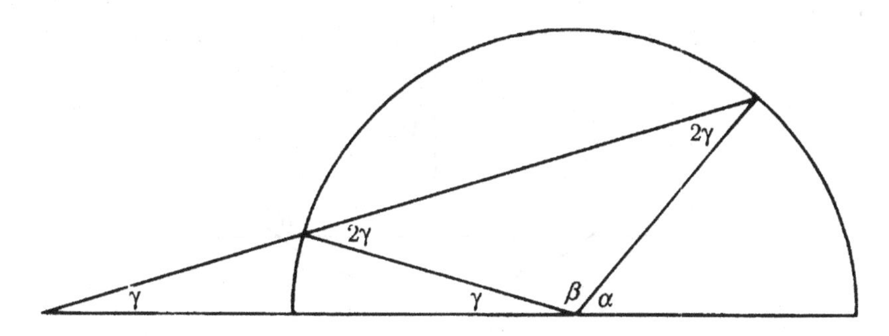

Fig. 1.2.7. Trisection

Exercises

1.2.1 Describe how to bisect a given angle with straightedge and compass.

1.2.2 Show that a quadratic equation with coefficients in the real quadratic closure of $\mathbb{Q}$ has its solutions in the real quadratic closure of $\mathbb{Q}$, if they are real.

1.3 The Constructible Numbers

We can express the constructibility of a point P in terms of numbers by using the *cartesian coordinates* of P, the distances from P to a fixed pair of perpendicular axes OX and OY (Figure 1.3.1). There is a straightedge and compass construction of the perpendicular M from a given point P to a given line L.

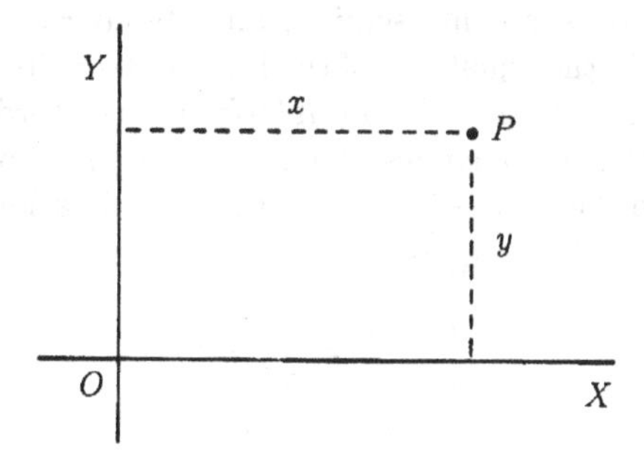

Fig. 1.3.1. Cartesian coordinates

For example, bisect the segment QR of L cut off by a suitable circle centered on P (Figure 1.3.2), using the method described in Section 1.2.

In particular, if O and the line OX are given we can construct OY, and

$P = (x, y)$ is constructible

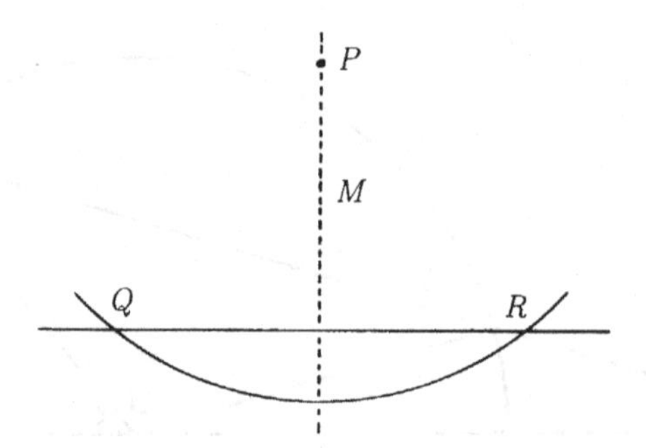

Fig. 1.3.2. Perpendicular bisector construction

$\Rightarrow$ the perpendiculars from P to OX, OY are constructible
$\Rightarrow x, y$ are constructible numbers.

Now we are ready for the characterisation of constructible numbers promised in Section 1.2.

Theorem. *A number d is constructible $\Leftrightarrow$ d belongs to the real quadratic closure of $\mathbb{Q}$.*

Proof. The proof of ($\Leftarrow$) is in Section 1.2.

To prove ($\Rightarrow$) it suffices to prove that the coordinates of any constructible point P are in the real quadratic closure of $\mathbb{Q}$, since the distance between constructible points (a_1, b_1), (a_2, b_2) is $\sqrt{(a_2 - a_1)^2 + (b_2 - b_1)^2}$ and hence in the real quadratic closure of $\mathbb{Q}$ as well.

If P is not one of the initial points, which we can take to be $(0, 0)$ and $(1, 0)$, then it is constructed as an intersection, and its coordinates are computable from the coefficients in the equations of the lines and/or circles being intersected. These coefficients, in turn, are computable from the coordinates of the points determining the lines and/or circles. Thus we have only to show that all these computations can be done by $+, -, \times, \div$ (rational operations) and $\sqrt{\ }$. Well, the line through (a_1, b_1) and (a_2, b_2) is

$$\frac{y - b_1}{x - a_1} = \frac{b_2 - b_1}{a_2 - a_1}$$

or

$$(b_1 - b_2)x + (a_2 - a_1)y = a_2 b_1 - b_2 a_1,$$

and the circle with center (a_1, b_1) and passing through (a_2, b_2) is

$$(x - a_1)^2 + (y - b_1)^2 = (a_2 - a_1)^2 + (b_2 - b_1)^2,$$

hence coefficients are computable rationally from coordinates. In the other direction, the coordinates of an intersection are computed from coefficients by solving linear or quadratic equations, hence by rational operations and square roots. □

Exercise

1.3.1 Find the intersections of the circle $(x - a_1)^2 + (y - b_1)^2 = r_1^2$ with the circle $(x - a_2)^2 + (y - b_2)^2 = r_2^2$ using $+, -, \times, \div$ and $\sqrt{}$. (Hint: First form the difference of the two equations.)

1.4 Some Famous Constructible Figures

Nearly everyone who has played with a compass has probably discovered the following Figure 1.4.1, which immediately gives constructions of the regular hexagon and the equilateral triangle. It is of course also easy to construct a square, but the construction of a regular pentagon is not at all obvious. A construction was discovered by the ancient Greeks. It can be based on the constructible number $\tau = (1+\sqrt{5})/2$, which happens to be the diagonal of a regular pentagon of side 1 (Figure 1.4.2).

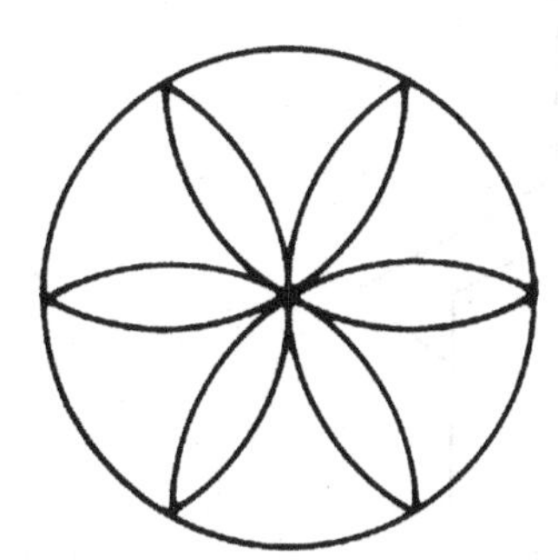

Fig. 1.4.1. Hexagon construction

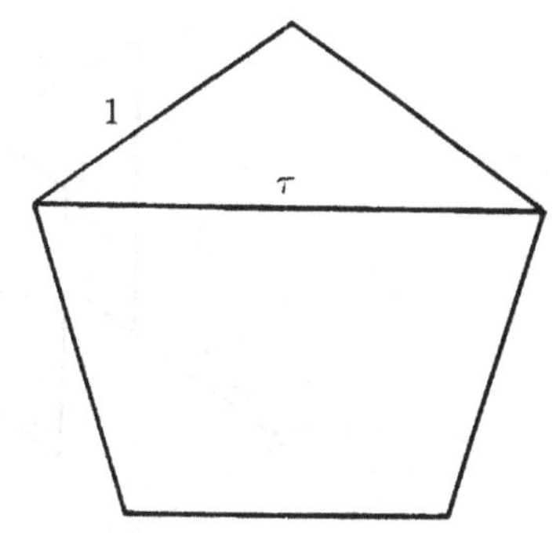

Fig. 1.4.2. Regular pentagon

To see why τ has this particular value, draw two more diagonals as shown in Figure 1.4.3. Each diagonal is parallel to the opposite side by symmetry, hence $ABDC$ is a parallelogram, hence $CD = AB = 1$. It follows that $CE = \tau - 1$, and, comparing ratios of corresponding sides in the similar triangles BCD and ECF, we get

$$\frac{1}{\tau} = \frac{\tau - 1}{1}.$$

This is equivalent to the equation $\tau^2 - \tau - 1 = 0$, the positive solution of which is the required value,

$$\tau = (1 + \sqrt{5})/2.$$

The number τ defines the so-called *golden rectangle* of height 1 and width τ (Figure 1.4.4). Its characteristic property is that, when a square is cut off, the rectangle that remains has the same shape as the original, namely

$$\frac{\text{short side}}{\text{long side}} = \frac{\tau - 1}{1} = \frac{1}{\tau}.$$

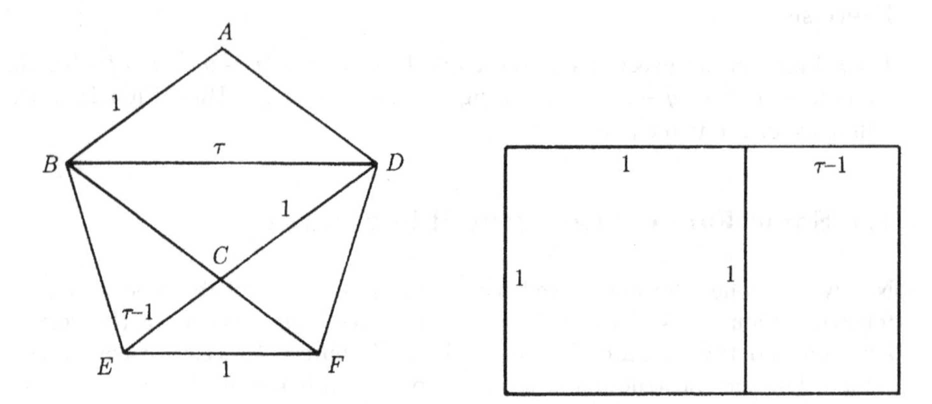

Fig. 1.4.3. Pentagon dimensions

Fig. 1.4.4. Golden rectangle

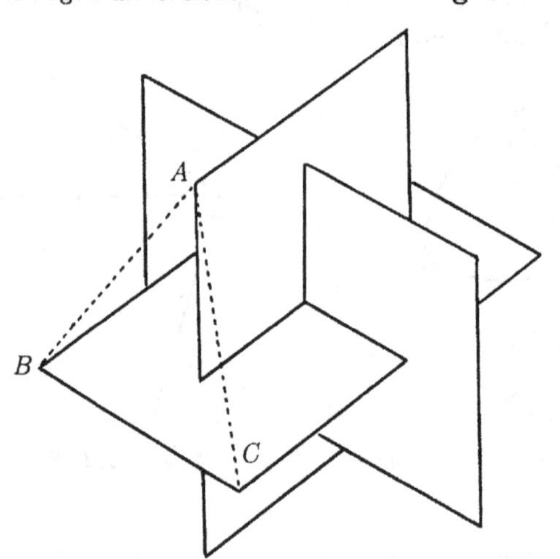

Fig. 1.4.5. Pacioli's construction of the icosahedron

Fitting three golden rectangles together as shown in Figure 1.4.5 yields the *icosahedron*, a solid figure bounded by 20 equilateral triangles, a typical one of which is ABC.

An easy application of the Pythagorean theorem shows that

$$AB = BC = 1 \quad \text{(Exercise 1.4.1).}$$

The symmetry of the figure implies that the diameter of the sphere circumscribing the icosahedron is just the diagonal of a golden rectangle, and hence constructible. (This construction of the icosahedron is due to Pacioli [1509]. Pacioli's book enthusiastically describes thirteen properties of τ, his favourite being its occurrence in the regular pentagon.)

One of the most beautiful discoveries of ancient Greek mathematics is that there are just five solids bounded by equal regular polygons. This follows from the fact that the angle in a regular n-gon is $(n-2)\pi/n$, which implies that $n = 3, 4$ or 5 for a bounding n-gon, since the sum of angles at a vertex must be less than 2π. Thus there are only five possible polygon combinations at a vertex : three, four or five 3-gons, three 4-gons or three 5-gons. The corresponding solids, called the *regular polyhedra*, are shown in Figure 1.4.6.

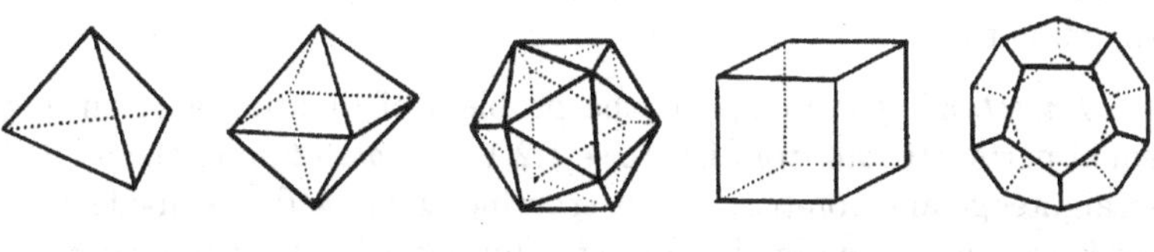

Fig. 1.4.6. Tetrahedron Octahedron Icosahedron Cube Dodecahedron

The regular polyhedra are studied systematically in Euclid's *Elements*, which culminates in a proof that for each of them the diameter of the circumscribing sphere is constructible relative to the side length of the polyhedron.

Exercises

1.4.1 Show that $AB = 1$ in Figure 1.4.5.

1.4.2 Show that the diagonal of a golden rectangle is $\sqrt{(5+\sqrt{5})/2}$.

1.4.3 Find tetrahedra whose vertices are vertices of the cube and dodecahedron.

1.4.4 By use of the cosine rule, show that $AB = \sqrt{2}$ in Figure 1.4.7.

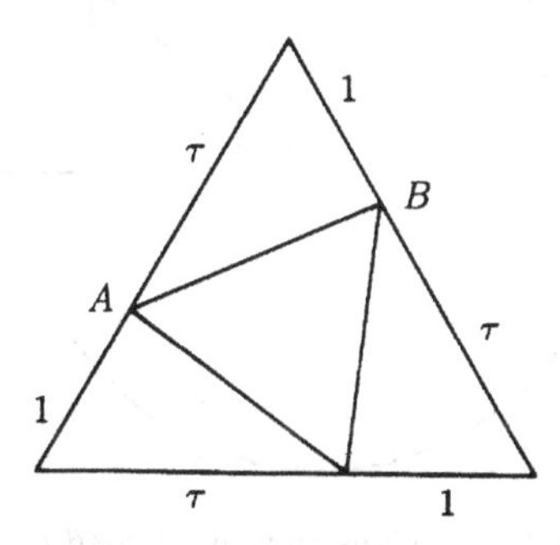

Fig. 1.4.7. Equilateral triangle with subdivided edges

1.4.5 Deduce from Exercise 1.4.4 that there is an icosahedron whose vertices lie on the edges of an octahedron, dividing them each in the ratio of τ to 1.

1.5 The Classical Construction Problems

Long before the algebraic nature of constructible numbers was understood, the Greeks attempted to solve problems equivalent to the solution of cubic or higher degree equations. The most famous of these are the following.

Duplication of the cube is the problem of constructing a cube with twice the volume of the unit cube. This amounts to solving the equation $x^3 = 2$, which gives the side x of the cube with volume equal to 2. (The special number 2 in this problem comes from a legend that the oracle at Delphi posed the problem of doubling a cubical altar. In general one has the problem of "multiplication of the cube by n.")

Trisection of the angle is the problem of dividing a given angle into three equal parts. (Recall from Exercise 1.2.1 that *bisection* of the angle is possible by straightedge and compass.) Constructing an angle θ is equivalent to constructing $\cos\theta$, as can be seen by describing a unit circle about the vertex of the angle and dropping a perpendicular as shown in Figure 1.5.1. Thus trisection is equivalent to constructing $\cos\theta$ when $\cos 3\theta$ is given. Since

$$\cos 3\theta = 4\cos^3\theta - 3\cos\theta$$

this finally is equivalent to solving

$$4x^3 - 3x = c$$

for an arbitrary value of $c = \cos 3\theta$. In particular, to trisect the (constructible) angle $\pi/3$ we have to solve this equation for $c = \cos(\pi/3) = 1/2$.

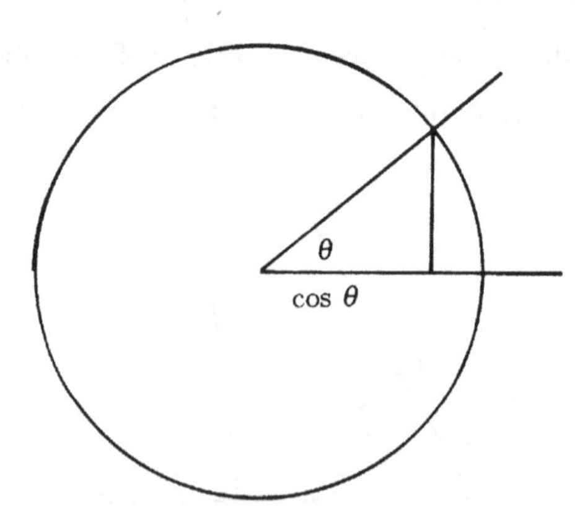

Fig. 1.5.1. Construction of $\cos\theta$ from θ

The problem of trisection generalises to n-section, the most interesting case of which is n-section of the unit circle, or *construction of the regular* n-*gon*. This problem is equivalent to construction of the point $(\cos\frac{2\pi}{n}, \sin\frac{2\pi}{n})$. By setting

$z = \cos\frac{2\pi}{n} + i\sin\frac{2\pi}{n}$ (see also Section 3.7) this point may also be described algebraically as a solution of $z^n = 1$. Since

$$z^n - 1 = (z-1)(z^{n-1} + z^{n-2} + \cdots + z + 1)$$

and we do not want the solution $z = 1$; the equation we really want to solve is

$$z^{n-1} + z^{n-2} + \cdots + z + 1 = 0.$$

Knowing that constructible numbers are obtained by (possibly nested) square roots, it is a reasonable guess that cubic problems, like duplication of the cube and trisection of the angle, are not solvable by straightedge and compass. However, the proof is still not obvious (see Chapter 5). In the case of the regular n-gon the answer is not even easy to guess; the n for which the regular n-gon is constructible are combinations of mysterious primes (see Chapters 5 and 9).

For the moment, the most important point to observe is that diverse geometric problems reduce to the solution of polynomial equations. Understanding polynomial equations is therefore a more fundamental problem, and of interest in its own right. In Section 1.6 we shall review the progress that was made on this problem after the ancient solution of quadratic equations.

Exercises

1.5.1 Explain how to construct $\sqrt[3]{2}$ as the intersection of a parabola with the hyperbola $xy = 1$.

1.5.2 Solve the equation $z^4 + z^3 + z^2 + z + 1 = 0$ for $z = \cos\frac{2\pi}{5} + i\sin\frac{2\pi}{5}$ by showing that the equivalent equation $z^2 + z + 1 + \frac{1}{z} + \frac{1}{z^2} = 0$ is quadratic in $y = z + \frac{1}{z}$. Deduce that the regular pentagon is constructible.

1.5.3 Similarly show that the equation $z^6 + z^5 + \cdots + z + 1 = 0$ for $z = \cos\frac{2\pi}{7} + i\sin\frac{2\pi}{7}$ is equivalent to a cubic in $y = z + \frac{1}{z}$.

1.5.4 Use your solution to Exercise 1.5.2 to find expressions for $\cos\frac{2\pi}{5}$ and $\sin\frac{2\pi}{5}$.

1.6 Quadratic and Cubic Equations

The usual method of solving the quadratic equation – *completing the square* – can also be viewed as "simplification by linear change of variable." The equation

$$ax^2 + bx + c = 0,$$

or equivalently

$$x^2 + \frac{b}{a}x + \frac{c}{a} = 0,$$

or equivalently

$$\left(x + \frac{b}{2a}\right)^2 - \frac{b^2}{4a^2} + \frac{c}{a} = 0,$$

is simplified by the substitution $y = x + \frac{b}{2a}$. The result is the equation

$$y^2 = \frac{b^2 - 4ac}{4a^2},$$

in which the first power of the variable is absent, so the solution is immediate on taking the square root of both sides. (Note that the expression $b^2 - 4ac$, called the *discriminant*, determines whether $ax^2 + bx + c$ is a perfect square. This happens just in case $b^2 - 4ac = 0$, in which case $x^2 + \frac{b}{a}x + \frac{c}{a} = \left(x + \frac{b}{2a}\right)^2$.)

In the case of the general cubic equation

$$ax^3 + bx^2 + cx + d = 0,$$

the analogous substitution $y = x + \frac{b}{3a}$ yields an equation of the form

$$y^3 = py + q,$$

where p, q are certain rational combinations of a, b, c, d. Thus the square of the variable is absent, but the equation is still not simple enough to solve immediately. The trick now is to substitute $u + v$ for y, obtaining

$$3uv(u+v) + u^3 + v^3 = p(u+v) + q.$$

The left-hand side $(u+v)^3$ has been given this particular arrangement in order to make it clear that the equation can be satisfied by setting

$$3uv = p, \quad u^3 + v^3 = q,$$

which we are free to do by choosing $v = p/3u$ and then solving

$$u^3 + \left(\frac{p}{3u}\right)^3 = q.$$

The latter equation is a quadratic in u^3, namely

$$(u^3)^2 - qu^3 + \left(\frac{p}{3}\right)^3 = 0,$$

so we *can* indeed solve it and the solutions for u^3 are

$$\frac{q}{2} \pm \sqrt{\left(\frac{q}{2}\right)^2 - \left(\frac{p}{3}\right)^3}$$

Since v^3 satisfies the same quadratic (notice the symmetric roles of u and v), we can assume without loss of generality that

$$u^3 = \frac{q}{2} + \sqrt{\left(\frac{q}{2}\right)^2 - \left(\frac{p}{3}\right)^3}, \quad v^3 = \frac{q}{2} - \sqrt{\left(\frac{q}{2}\right)^2 - \left(\frac{p}{3}\right)^3},$$

whence

$$y = u + v = \sqrt[3]{\frac{q}{2} + \sqrt{\left(\frac{q}{2}\right)^2 - \left(\frac{p}{3}\right)^3}} + \sqrt[3]{\frac{q}{2} - \sqrt{\left(\frac{q}{2}\right)^2 - \left(\frac{p}{3}\right)^3}}$$

is the solution of $y^3 = py + q$.

Exercises

1.6.1 Show that the substitution $y = x + \frac{a_{n-1}}{na_n}$ in $a_n x^n + \ldots + a_1 x + a_0 = 0$ yields an equation in which the y^{n-1} term is absent.

1.6.2 (Bombelli [1572]) Reconcile the solution $y = \sqrt[3]{2 + 11i} + \sqrt[3]{2 - 11i}$ of $y^3 = 15y + 4$ with the obvious solution $y = 4$ by showing $(2 \pm i)^3 = 2 \pm 11i$.

1.6.3 (Viète [1591]) Show that $y^3 - py = q$ reduces to the form $4z^3 - 3z = c$ by a suitable substitution $y = mz$, whence one has a "solution by trisection" by finding $z = \cos\theta$ where $c = \cos 3\theta$.

1.6.4 Compare the role of $(q/2)^2 - (p/3)^3$ in $y^3 = py + q$ with the role of $b^2 - 4ac$ in $ax^2 + bx + c = 0$.

1.7 Quartic Equations

To solve the general quartic equation

$$ax^4 + bx^3 + cx^2 + dx + e = 0$$

we begin with the substitution $y = x + \frac{b}{4a}$, which yields an equation of the form

$$y^4 + py^2 + qy + r = 0,$$

or equivalently

$$(y^2 + p)^2 = py^2 - qy + p^2 - r.$$

Then for any z,

$$\begin{aligned}(y^2 + p + z)^2 &= (py^2 - qy + p^2 - r) + 2z(y^2 + p) + z^2 \\ &= (p + 2z)y^2 - qy + (p^2 - r + 2pz + z^2).\end{aligned}$$

We can make the quadratic $Ay^2 + By + C$ on the right-hand side a perfect square by choosing z so that $B^2 - 4AC = 0$, which is a cubic equation for z.

Thus z can be found by the method described in Section 1.6. We can then take the square root of both sides, and finally solve the resulting quadratic equation for y.

Exercise

1.7.1 Show that solving $y^4 + py^2 + qy + r = 0$ is equivalent to finding the intersection of the parabola $y^2 = x$ with another parabola.

1.8 Solution by Radicals

The common feature of the solutions of quadratic, cubic and quartic equations is that they all express the solution in terms of the coefficients by means of $+, -, \times, \div$ and m^{th} roots, in fact just square roots and cube roots. In general, an equation

$$a_n x^n + a_{n-1}x^{n-1} + \cdots + a_1 x + a_0 = 0$$

is said to be *solvable by radicals* if the solution is expressible in terms of $a_0, \ldots, a_n$ by means of the rational operations $+, -, \times, \div$ and m^{th} roots $\sqrt{\ }, \sqrt[3]{\ }, \sqrt[4]{\ }, \ldots$.

After the solution of cubic and quartic equations by radicals in the 16th century, the main goal of algebra was the solution of the general quintic (fifth degree) equation. There were so many unsuccessful attempts that eventually mathematicians turned to the more profound problem of proving that a solution of the general quintic by radicals is *impossible*. The proofs of this fact by Abel [1826] and Galois [1831] were a quantum leap in the development of algebra, a leap above the details of computation to a realm of powerful abstract concepts. The power of these abstract concepts – groups, fields and dimension – lies in their ability to capture general features of computation, so that the existence of particular computations can be proved or disproved without attempting to carry them out.

In the chapters that follow we shall develop these concepts from their roots in elementary arithmetic, so that their *naturalness* may be seen as well as their power. The concepts of field and dimension suffice, in fact, to settle the classical construction problems, as we shall see in Chapter 5. The concept of group is crucial in the proof that the general quintic equation is unsolvable, as we shall see in Chapter 8. It is also very helpful in determining which regular n-gons are constructible, as we shall see in Chapter 9.

Exercises

1.8.1 Assuming the de Moivre theorem $\cos n\theta \pm i \sin n\theta = (\cos\theta \pm i \sin\theta)^n$, where $i^2 = -1$, show that $\cos n\theta$ is a polynomial $p(\cos\theta)$ in $\cos\theta$.

1.8.2 (de Moivre [1707]) Show that the equation $p(x) - y = 0$ relating $x = \cos\theta$ to $y = \cos n\theta$ has the solution by radicals

$$x = \frac{1}{2}\sqrt[n]{y + \sqrt{y^2 - 1}} + \frac{1}{2}\sqrt[n]{y - \sqrt{y^2 - 1}}$$

(Thus "n-section of the angle" is solvable by radicals.)

1.9 Discussion

The straightedge and compass have been the trademark of geometry since the appearance of Euclid's *Elements* around 300 BC. In the *Elements*, Euclid set the agenda and style of mathematics for the next 2000 years, and even the most radical new advances – theory of equations in the 16th century, analytic geometry and calculus in the 17th, and abstract algebra in the 19th – were in some sense an extension of Euclid. In the case of algebra, this should become clear as the present book unfolds. The most familiar aspect of the *Elements'* style is the axiomatic method, the "definition – theorem – proof" format. However, almost equally important is the algorithmic content, the "constructions," which are almost as common as the theorems. Sometimes a construction is part of the proof of a theorem, but often the construction stands alone: the whole point being to prove that a certain figure can be constructed with straightedge and compass. Indeed, the *Elements* begins with the construction of the equilateral triangle and ends with the construction of the five regular solids.

Euclid was aware that straightedge and compass construction involves rational operations and square roots, but he stopped short of describing the real quadratic closure of $\mathbb{Q}$. Book X of the *Elements* contains a thorough discussion of quantities of the form $a \pm b\sqrt{c}$, $\sqrt{a \pm b\sqrt{c}}$ and $\sqrt{\sqrt{a} \pm \sqrt{c}}$ where $a, b, c \in \mathbb{Q}$, but not of any deeper nesting of square roots, since this is as far as one needs to go to construct the regular solids. For example, the side of an icosahedron inscribed in the unit sphere is $\frac{1}{5}\sqrt{10(5-\sqrt{5})}$ (see Heath [1925] vol. III, p. 489 and Exercise 1.9.2 below).

Why the restriction to straightedge and compass? Probably for the sake of simplicity, the same reason Euclid restricted his plane geometry to the properties of straight lines and circles. The restriction was not observed universally in Greek mathematics. Constructions using other curves were sometimes used in cases where straightedge and compass failed, as with the duplication of the cube and trisection of the angle. An interesting account of some of these constructions, and their influence on the modern theory of curves, may be found in Brieskorn and Knörrer [1986].

New light was thrown on construction problems with the development of algebra in the 16th century. Around 1500, in Bologna, del Ferro found the solution to the cubic equation. The solution was rediscovered by Tartaglia in the 1530s, and published in Cardano's *Ars Magna* [1545]. This book also gave the solution to the quartic equation, which was found by Cardano's student Ferrari. Complementing these Italian discoveries, the French mathematician Viète [1591] found that the solution of cubic equations is equivalent to trisection (compare with Exercise 1.6.3). Thus, by the beginning of the 17th century, algebra had matured to the point where it could become an equal partner with geometry.

The partnership eventuated in the (independent) work of Fermat [1629] and Descartes [1637] on algebraic geometry. By representing curves by their now familiar "cartesian" equations, Fermat and Descartes were able to use algebra to solve geometric problems and vice versa. They took advantage of the ease and freedom with which curves could be defined by equations to extend the range of constructions, intersecting 2nd and 3rd degree curves to find solutions of $4^{\text{th}}, 5^{\text{th}}$ and 6th degree equations. This idea clarified and systematised some of the constructions found by the Greeks. For example, a duplication of the cube found by Menaechmus (around 350 BC) could be viewed as solving $x^3 = 2$ by finding the intersection of $y = \frac{1}{2}x^2$ and $y = \frac{1}{x}$ (compare with Exercise 1.5.1). However, the general notion of construction proved to be unwieldy and was abandoned around 1750. By that time it seemed clear that the solution of equations was best left to algebra, and that the theory of constructions was best confined to Euclid's geometry, where it had a natural place. Algebraic geometry could stand on its own feet as a theory of curves.

Then in 1796 the 19-year old Gauss put the shine back on the straightedge and compass with an astonishing new construction – the regular 17-gon. This was the first new regular polygon constructed since ancient times, and incidentally a big factor in Gauss's decision to become a mathematician. It was also one of the key events in the development of modern algebra, involving insight into the hidden, abstract properties of polynomials just as much as manipulative skill. Gauss expanded his discovery to a general theory of regular n-gons in his classic *Disquisitiones Arithmeticae* [1801]. Over the next few decades several such insights were gained by Gauss and his brilliant younger contemporaries Abel and Galois, as we shall see in later chapters. Between them they solved most of the classical problems of construction by straightedge and compass, and solution of equations by radicals. Strangely enough, the impossibility of duplication of the cube and trisection of the angle was first proved by an obscure mathematician, Wantzel [1837]. Wantzel also filled a gap in Gauss' theory of regular n-gons (see Chapter 5), and gave an explicit algebraic criterion for constructibility, previously only implicit in Descartes [1637].

Perhaps the most famous construction problem was settled by Lindemann [1882]. He proved the impossibility of "squaring the circle," the construction of a square equal in area to a given circle. This amounts to construction of the number π. We have omitted mention of squaring the circle until now because it is not in fact an algebraic problem. Lindemann showed that π is not the solution of any polynomial equation with integer coefficients. As we shall see in Chapter 5, this means that π is not only not constructible, it cannot be expressed as any finite combination of rational operations and n^{th} roots.

In view of this result, it now seems fair to say that the Greeks were hopelessly out of their depth in attempting to square the circle. At the time, however, they had reason to suppose they could succeed. Hippocrates of Chios (around 450 BC) made the remarkable discovery that certain regions bounded by circular arcs *are* squarable. The regions in question are called *lunes* because their shape is like that of a crescent moon. The general lune is bounded by two circular arcs, subtending angles $2\theta, 2\theta'$ as shown in Figure 1.9.1. The simplest squarable case is where $\theta = \pi/2, \theta' = \pi/4$ (Exercise 1.9.1).

Hippocrates discovered this, as well as two other squarable lunes, for which $\theta/\theta' = 3, 3/2$. Two more squarable lunes were found by Wallenius [1766], with $\theta/\theta' = 5, 5/3$. They were rediscovered by Clausen [1840], who conjectured that

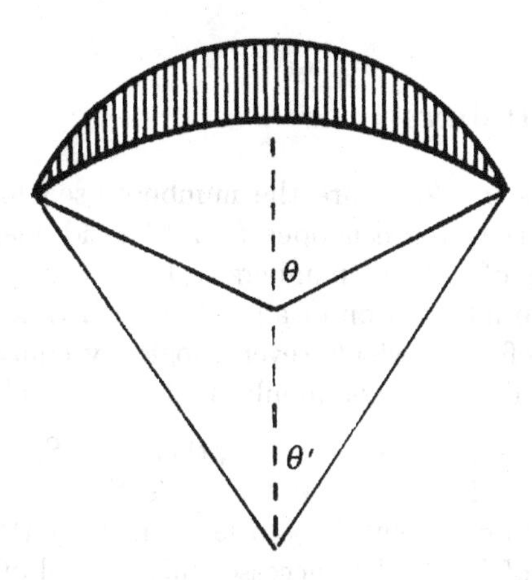

Fig. 1.9.1. A lune

the five squarable lunes then known were the only ones with θ/θ' rational. This turns out to be correct, and the proof is a relatively recent application of Galois theory (Čebotarev [1934], completed by Dorodnov [1947]).

I suppose this means that squarable lunes are a fluke, but an interesting fluke nevertheless. Perhaps they could even be ranked with the other two "famous fives" of constructibility theory: the five regular polyhedra and the five known primes p for which the regular p-gon is constructible, namely $p = 3, 5, 17, 257, 65537$ (see Chapters 5 and 9).

Exercises

1.9.1 Compute the area of the lune with $\theta = \pi/2, \theta' = \pi/4$.

1.9.2 Deduce from Exercise 1.4.2 that the side of an icosahedron inscribed in the unit sphere is $\frac{1}{5}\sqrt{10(5-\sqrt{5})}$.

2 The Rational Numbers

2.1 Natural Numbers

The natural numbers $0, 1, 2, \ldots$ are the numbers used for counting. They are generated from 0 by the *successor* operation $+1$ (add one). In other words, the set $\mathbb{N} = \{0, 1, 2, \ldots\}$ of natural numbers is the closure of the set $\{0\}$ under successor, that is, the intersection of all sets S such that $0 \in S$ and $n+1 \in S$ when $n \in S$. This definition yields several logically equivalent properties of $\mathbb{N}$ known as *induction*. The most commonly used versions of induction are:

I. If $0 \in S$, and if $n+1 \in S$ when $n \in S$, then $\mathbb{N} \subseteq S$.
II. If $0 \in S$, and if $n+1 \in S$ when $0, 1, \ldots, n \in S$, then $\mathbb{N} \subseteq S$.
III. If $T \subseteq \mathbb{N}$ is nonempty then T has a least member (that is, an $n \in T$ such that the closure of $\{n\}$ under successor includes all of T).

We show these statements are equivalent by proving I $\Rightarrow$ II $\Rightarrow$ III $\Rightarrow$ I.

(I $\Rightarrow$ II) is probably the trickiest. Suppose S satisfies the assumptions of II, that is, $0 \in S$, and $n+1 \in S$ whenever $0, \ldots, n \in S$. To prove that $\mathbb{N} \subseteq S$ we consider

$$S' = \{n : 0, \ldots, n \in S\}$$

and prove instead that $\mathbb{N} \subseteq S'$. Since $0 \in S$ by I, we have $0 \in S'$. Moreover,

$$\begin{aligned} n \in S' &\Rightarrow 0, \ldots, n \in S \\ &\Rightarrow n+1 \in S \text{ by the assumptions of II} \\ &\Rightarrow 0, \ldots, n+1 \in S \\ &\Rightarrow n+1 \in S'. \end{aligned}$$

Thus $\mathbb{N} \subseteq S'$ by I and, since $S' \subseteq S, \mathbb{N} \subseteq S$.

(II $\Rightarrow$ III). Suppose $T \subseteq \mathbb{N}$ and let $S = \mathbb{N} - T$. If T has no least member then $0 \in S$, otherwise 0 is the least member of T. Also

$$\begin{aligned} 0, \ldots, n \in S &\Rightarrow 0, \ldots, n \notin T \\ &\Rightarrow n+1 \notin T \text{ otherwise } n+1 \text{ is the least member of } T \\ &\Rightarrow n+1 \in S. \end{aligned}$$

Hence, by II, $\mathbb{N} \subseteq S$ and therefore T is empty.

(III $\Rightarrow$ I). Suppose $0 \in S$, and $n+1 \in S$ whenever $n \in S$. Let $T = \mathbb{N} - S$. Then $0 \notin T$ since $0 \in S$ and

$$\begin{aligned} 0 < n \in T &\Rightarrow n \notin S \\ &\Rightarrow n-1 \notin S \text{ by the assumptions of I} \\ &\Rightarrow n-1 \in T. \end{aligned}$$

Hence T has no least member and therefore $T = \mathbb{N} - S$ is empty by III. Thus $\mathbb{N} \subseteq S$. $\square$

We make use of induction in both definitions and proofs. An example of *definition by induction* is the following definition of the function +.

$$m + 0 = m \quad \text{for all } m \in \mathbb{N} \tag{0}$$

$$m + (k+1) = (m+k) + 1 \quad \text{for all } m, k \in \mathbb{N} \tag{$k+1$}$$

The clause (0) defines $m + n$ for $n = 0$ (and all $m \in \mathbb{N}$). Then clause $(k+1)$ defines $m + n$ for $n = k + 1$ whenever $m + n$ is already defined for $n = k$. Thus the set of n for which $m + n$ is defined includes 0 and is closed under successor, hence it includes all $n \in \mathbb{N}$ by version I of induction. Thus $m + n$ is defined not only for all m, but also for all n, in $\mathbb{N}$.

Properties P of functions defined by induction are naturally *proved by induction*, that is, by proving that the set of n for which P holds includes 0 and is closed under successor (hence is all of $\mathbb{N}$). For example, we can prove the *associative* property of +,

$$l + (m + n) = (l + m) + n$$

by *induction on n*, showing associativity holds for $n = 0$ (the "base step") and that it holds for $n = k + 1$ whenever it holds for $n = k$ (the "induction step").

Base step, $n = 0$:

$$\begin{aligned} l + (m+0) &= l + m \quad \text{since} \quad m + 0 = m \quad \text{by definition of } + \\ &= (l + m) + 0 \quad \text{by definition of } +. \end{aligned}$$

Hence associativity holds for $n = 0$.

Induction step, $n = k + 1$:

$$\begin{aligned} l + (m + (k+1)) &= l + ((m+k) + 1) \\ &\qquad \text{since} \quad m + (k+1) = (m+k) + 1 \quad \text{by definition of } + \\ &= (l + (m+k)) + 1 \quad \text{by definition of } + \\ &= ((l+m) + k) + 1 \quad \text{if associativity holds for } n = k \\ &= (l+m) + (k+1) \quad \text{by definition of } +. \end{aligned}$$

Hence associativity holds for $n = k+1$ whenever it holds for $n = k$, and therefore it holds for all n. □

The basic properties of multiplication can also be proved by induction, starting with the inductive definition $m \times 0 = 0, m(n+1) = mn + m$. In particular, one has the associative property of multiplication, $l(mn) = (lm)n$, which means that brackets are unnecessary for products as well as sums.

Exercises

2.1.1 Prove $0 + m = m$ by induction on m.

2.1.2 Prove $m + n = n + m$ (commutative property of +) by induction on n.

2.1.3 Prove $l(m + n) = lm + ln$ (distributive property) by induction on n.

2.1.4 Define m^n by induction.

2.1.5 The *Fibonacci numbers* F_n are defined inductively by $F_0 = 0, F_1 = 1$ and $F_{k+2} = F_{k+1}+F_k$. Prove by induction that $1+1/(1+1/(1+\cdots 1)\cdots) = F_{n+1}/F_n$.

2.1.6 Prove by induction that $\tau^n = \tau F_n + F_{n-1}$ for each root $\tau_+, \tau_- = (1 \pm \sqrt{5})/2$ of the equation $\tau^2 = \tau + 1$, and hence show that $F_n = (\tau_+^n - \tau_-^n)/(\tau_+ - \tau_-)$.

2.2 Integers and Rational Numbers

The natural numbers are closed under addition and multiplication but not under subtraction (you can't take 7 from 3...). To overcome this inconvenience we enlarge $\mathbb{N}$ to the set $\mathbb{Z}$ of *integers* by introducing a *negative integer* $-n$ for each nonzero $n \in \mathbb{N}$ (now renamed the set of *positive integers*). This element is called the *additive inverse* of n because $n + (-n) = 0$. The definitions of addition, multiplication and additive inverse are extended to $\mathbb{Z}$ so that the following properties of natural numbers hold for all integers:

$$m + n = n + m, \quad mn = nm \qquad (\textit{commutative})$$

$$l + (m + n) = (l + m) + n, \quad l(mn) = (lm)n \qquad (\textit{associative})$$

$$l(m + n) = lm + ln \qquad (\textit{distributive})$$

$$n + 0 = n, \quad n \times 1 = n \qquad (\textit{identity})$$

$$n + (-n) = 0 \qquad (\textit{additive inverse})$$

These properties define what is called a *commutative ring with unit* (the unit being 1 in this case). Since this is the only type of ring we shall consider in this book, we shall simply call $\mathbb{Z}$ a ring and refer to the above properties of $\mathbb{Z}$ as its *ring* properties.

Subtraction is defined for all $m, n \in \mathbb{Z}$ by

$$m - n = m + (-n),$$

and the properties of subtraction then follow from the ring properties of $\mathbb{Z}$. In particular, one finds that the additive inverse of n is unique, that $-(-n) = n$, and that $(-1)(-1) = 1$ (see Exercise 2.2.1).

In principle, the negative integers are unnecessary, since all statements involving them can be replaced by equivalent statements about natural numbers. For example, $3-7 = -4$ is equivalent to $3-7+4 = 0$. But in practice they confer an enormous advantage – the advantage of algebra over arithmetic – by allowing the properties of subtraction to be used without restriction. For example, in $\mathbb{Z}$ one can say

$$(m - n)(m + n) = m^2 - n^2$$

without adding "provided $m \geq n$," as one must when working in $\mathbb{N}$.

Since $\mathbb{Z}$ is itself not closed under division, it is similarly convenient to enlarge it to the set of *rational numbers* $\mathbb{Q}$ by introducing a *multiplicative inverse* n^{-1} for each nonzero $n \in \mathbb{Z}$, with $nn^{-1} = 1$, and closing under multiplication. The definitions of addition and multiplication are extended to the resulting expressions $mn^{-1} = m/n$ by

$$\frac{m_1}{n_1} + \frac{m_2}{n_2} = \frac{m_1 n_2 + m_2 n_1}{n_1 n_2}$$

and

$$\frac{m_1}{n_1}\frac{m_2}{n_2} = \frac{m_1 m_2}{n_1 n_2}$$

(the usual rules for adding and multiplying fractions). Then $\mathbb{Q}$ has all the ring properties of $\mathbb{Z}$ and, since m/n has the multiplicative inverse n/m, we have an r^{-1} for each nonzero $r \in \mathbb{Q}$ with

$$rr^{-1} = 1 \qquad \textit{(multiplicative inverse)}$$

A ring with this additional property is called a *field.*

Division is defined for all $r, s \in \mathbb{Q}$, $s \neq 0$, by $r \div s = rs^{-1}$. Thus $\mathbb{Q}$ is closed under $+, -, \times, \div$ (by nonzero elements). This is the ideal situation for computation, and indeed the field properties facilitate computation so greatly that they enable us to see things that are virtually invisible from the viewpoint of $\mathbb{N}$. For example, the ability to solve linear equations in any number of variables leads to the invaluable concepts of vector space and dimension (see Section 5.3).

Nevertheless, the rational numbers have no content not already implicit in the natural numbers, and one is often forced back to $\mathbb{N}$ in order to answer questions about $\mathbb{Q}$. For example, the field properties of $\mathbb{Q}$ are no help in deciding whether there is $x \in \mathbb{Q}$ such that $x^2 = 2$. The only way to decide is by attacking the equivalent question about $\mathbb{N}$: are there $m, n \in \mathbb{N}$ such that $m^2 = 2n^2$? (See Section 3.1.) In Section 2.3 we return to $\mathbb{N}$ in order to develop an approach to such questions.

Exercises

2.2.1 Prove

(i) $l + m = l + n \Rightarrow m = n$

(ii) uniqueness of the additive inverse

(iii) $-(-n) = n$

(iv) $(-1)n = -n$

(v) $(-1)(-1) = 1$.

2.2.2 Does $\mathbb{Q}$ contain any commutative ring with unit, other than $\mathbb{Z}$?

2.3 Divisibility

The fact that $\mathbb{N}$ is *not* closed under division makes the problem of divisibility in $\mathbb{N}$ an interesting problem. In fact, it is the source of some of the most challenging problems in mathematics.

We say that $a \in \mathbb{Z}$ is *divisible by* $b \in \mathbb{Z}$ (or that a is a *multiple of* b, or that b is a *divisor of* a, or that b *divides* a) if

$$a = bc \quad \text{for some} \quad c \in \mathbb{Z}.$$

We sometimes abbreviate this relation by $b|a$, and also write $b \nmid a$ when b does not divide a. It is immediate that any $n \in \mathbb{N}$ has divisors 1 and n, so we call these the *trivial* divisors of n. Divisors are sometimes called *factors*, particularly in the case where a is written as a product $a = b_1 b_2 \cdots b_k$. The b_is are called factors, and their product a *factorisation*, of a. A natural number $p \neq 1$ with no nontrivial divisors is called a *prime number* or simply a *prime*. We exclude 1 from the primes because it has exceptional properties which would spoil certain general statements, in particular unique prime factorisation (see Section 2.5).

Two easy consequences of the definition of divisor are: *a divisor of a divisor is a divisor*, that is, if $b|a$ and $c|b$ then $c|a$, and *every natural number $n > 1$ has a prime divisor.*

The first assertion is immediate from the definition because

$$\begin{aligned} b|a &\Rightarrow a = be \quad \text{for some} \quad e \in \mathbb{N} \\ c|b &\Rightarrow b = cd \quad \text{for some} \quad d \in \mathbb{N} \\ &\Rightarrow a = cde \\ &\Rightarrow c|a. \end{aligned}$$

And the second assertion follows from the first by choosing a nontrivial divisor m of n, if one exists, then a nontrivial divisor l of m, and so on. Since $n > m > l > \ldots$ there must be a least divisor p in the sequence, by version III of induction (Section 2.1). Then p is prime, since it has no nontrivial divisor, and p is a divisor of n by the first assertion.

The existence of prime divisors yields the first important theorem about primes.

Euclid's Theorem. *There are infinitely many primes.*

Proof. Given any primes $p_1, \ldots, p_k$ we can find another prime as follows. Consider the natural number

$$n = p_1 \cdots p_k + 1.$$

Then none of $p_1, \ldots, p_k$ divide n – they all leave remainder 1 – hence any prime divisor of n is a prime $\neq p_1, \ldots, p_k$. □

Another consequence of prime divisors is a *prime factorisation* for each natural number n. If q_1 is a prime divisor of n, then n/q_1 is a smaller natural number with another prime divisor q_2 and so on, hence (by another application of version III of induction)

$$n = q_1 \cdots q_s$$

for some primes $q_1, \ldots, q_s$. What is not clear, at this stage, is whether n is divisible by any prime $\neq q_1, \ldots, q_s$ (see Section 2.5).

We have developed the fundamentals of divisibility in $\mathbb{N}$ rather than $\mathbb{Z}$ because primes are simpler to define in $\mathbb{N}$. The definition of divisibility obviously applies equally well to $\mathbb{Z}$. Likewise, one can define a prime in $\mathbb{Z}$ to be any integer with no nontrivial divisors; it is just that the trivial divisors of $n \in \mathbb{Z}$ are $\pm 1, \pm n$.

In $\mathbb{N}$ we can say that a nontrivial divisor m of n satisfies $1 < m < n$, as we did (implicitly) in proving that n has a prime divisor. In $\mathbb{Z}$ we have to say $1 < |m| < |n|$, where $|\ |$ is the *absolute value* function defined by

$$|m| = \begin{cases} m & \text{if } m \geq 0 \\ -m & \text{if } m < 0 \end{cases}$$

Exercises

2.3.1 Show that if $c|a$ and $c|b$ then $c|(a+b)$, and note the role of the distributive property in the proof.

2.3.2 Show that $2^n - 1$ is prime only if n is prime. Is the converse true?

2.4 The Euclidean Algorithm

Finding prime divisors of a natural number n is simple in principle – just try dividing n by all smaller numbers – but in practice it is almost as laborious as counting up to n. It is not feasible to do it for a number n with, say, 1000 digits. It turns out to be much easier to find the *greatest common divisor* $\gcd(a, b)$ of two integers a,b. Moreover, the method by which this is done – the *Euclidean algorithm* – ultimately yields a better understanding of primes, not only in $\mathbb{Z}$, but also in other algebraic domains with a concept of "divisor."

The fundamental property of common divisors is that if $c|a$ and $c|b$ then $c|(ma+nb)$ for any $m, n \in \mathbb{Z}$. The Euclidean algorithm is based on this property, together with the following property of natural numbers which may be called the *division property* of $\mathbb{N}$: for any $a, b \in \mathbb{N}$ with $b \neq 0$ there are $q, r \in \mathbb{N}$ such that

$$a = qb + r \quad \text{and} \quad 0 \leq r < b$$

We call q the *quotient*, and r the *remainder of a on division by b*.

A proof of the division property, which also gives a way to find r, goes as follows. If $b > a$ then $r = a$ (and $q = 0$). If $a > b$ then subtract b from a repeatedly until a number $r < b$ is obtained. We must reach such a number because otherwise $a - b, a - 2b, a - 3b, \ldots$ form a set of natural numbers without least member, contrary to version III of induction. □

With this algorithm for remainders in hand, we are ready for the *Euclidean algorithm on natural numbers* $a, b > 0$. It begins by setting $a_1 = \max(a, b), b_1 = \min(a, b)$, in general computes

$$a_{k+1} = b_k$$
$$b_{k+1} = r_k = a_k - q_k b_k \text{ where } 0 \leq r_k < b_k \text{ (remainder on division by } b_k\text{)},$$

and halts when $r_k = 0$.

Theorem. *The last nonzero remainder* $r_l = b_{l+1}$ *produced by the Euclidean algorithm is* $\gcd(a, b)$, *and*

$$\gcd(a, b) = ma + nb \text{ for some } m, n \in \mathbb{Z}.$$

Proof. Since any common divisor of a_k, b_k is also a divisor of $a_{k+1} = b_k$ and $b_{k+1} = a_k - q_k b_k$ (by the fundamental property of common divisors), we have

$$\gcd(a_1, b_1) = \gcd(a_2, b_2) = \cdots = \gcd(a_{l+1}, b_{l+1}).$$

And $\gcd(a_{l+1}, b_{l+1}) = b_{l+1}$ because a_{l+1} has zero remainder r_{l+1} on division by b_{l+1}.

Thus

$$\gcd(a, b) = \gcd(a_{l+1}, b_{l+1}) = b_{l+1}.$$

Also, each $a_{k+1} = b_k$ and $b_{k+1} = a_k - q_k b_k$ is a linear combination of a_k, b_k with integer coefficients. It follows by induction on k that a_{k+1} and b_{k+1} are linear combinations of $a_1 = a, b_1 = b$ with integer coefficients, and in particular this is true of $b_{l+1} = \gcd(a, b)$. □

There is no need to generalise the Euclidean algorithm to negative integers since obviously $\gcd(a, b) = \gcd(\pm a, \pm b)$, and hence one may always arrange to work with natural numbers. However, the coefficients m, n such that $\gcd(a, b) = ma + nb$ certainly can be negative, hence $\mathbb{Z}$ is a more appropriate place to discuss greatest common divisors.

Exercises

2.4.1 Show that the gcd may also be obtained by a "subtractive" form of the Euclidean algorithm which computes

$$a_{k+1} = \max(a_k, b_k) - \min(a_k, b_k)$$
$$b_{k+1} = \min(a_k, b_k)$$

until $a_l = b_l$.

2.4.2 Find $\gcd(15, 28)$ in the form $15m + 28n$.

2.4.3 Use the Euclidean algorithm to show that $\gcd(F_{n+1}, F_n) = 1$, where F_n is the nth Fibonacci number. Also find l, m such that $lF_n + mF_{n+1} = 1$.

2.4.4 Show that an equation $ax + by = c$, where $a, b, c \in \mathbb{Z}$, has a solution $x, y \in \mathbb{Z} \Leftrightarrow \gcd(a, b) | c$.

2.4.5 Use $\gcd(p_1 p_2 \cdots p_n, p_1 p_2 \cdots p_n - 1)$ to give another proof that $p_1, p_2, \ldots, p_n$ are not all the primes.

2.5 Unique Prime Factorisation

The gcd is relevant to the study of prime numbers because $\gcd(a,p)$ is p or 1 when p is prime, since $1, p$ are the only natural number divisors of p. Hence, *if a is not a multiple of the prime p then* $\gcd(a,p) = 1$. When combined with the representation of the gcd obtained from the Euclidean algorithm, this simple observation about primes leads to the following:

Prime Divisor Lemma. *If p is prime and $p|ab$ then $p|a$ or $p|b$.*

Proof. Suppose that a is not a multiple of p, so that

$$1 = \gcd(a,p) = ma + np \quad \text{for some} \quad m, n \in \mathbb{Z}.$$

Multiplying both sides by b we get

$$b = mab + npb.$$

Now p divides ab by hypothesis, and p plainly divides npb, thus p divides both terms on the right-hand side. Hence p divides the left-hand side, b. □

It follows by repeated application of this lemma that if $p|q_1 \ldots q_k$ then $p|q_1$ or $\ldots$ $p|q_k$. The lemma goes back to Euclid's *Elements*, Book VII, 30 but the following consequence (nowadays considered to be more important) first appeared in Gauss [1801], article 16. It is often called the *fundamental theorem of arithmetic.*

Unique Prime Factorisation Theorem. *The factorisation of each natural number into primes is unique (up to the order of factors).*

Proof. Suppose $p_1 \cdots p_r = q_1 \cdots q_s$ are two prime factorisations of the same number. Since the prime p_1 divides the left-hand side it also divides $q_1 \ldots q_s$, and hence $p_1|q_1$ or $\ldots$ or $p_1|q_s$ by the lemma. Since one prime divides another only if they are identical, it follows that $p_1 = q_i$ for some i.

We can therefore cancel $p_1 = q_i$ from each side and repeat the argument. Eventually the last factor is cancelled from one side, and the last factor from the other side must be cancelled with it, since a product of primes cannot equal 1. Hence the prime factors $p_1, \ldots, p_r$ are the same (up to order) as the prime factors $q_1, \ldots, q_s$. □

We can now see why the number 1 is excluded from the primes by definition. Factorisations involving 1 are *not* unique, for example $2 = 1 \times 2$, hence unique prime factorisation fails if we include 1 among the primes.

Exercises

2.5.1 Suppose $n = p_1^{e_1} \cdots p_r^{e_r}$ where $p_1, \ldots, p_r$ are the *distinct* primes that divide n. Show that $m|n \Leftrightarrow m = p_1^{d_1} \cdots p_r^{d_r}$, where each $d_i \leq e_i$.

2.5.2 Deduce formulas for $\gcd(a,b)$ and $\operatorname{lcm}(a,b)$ in terms of the prime powers in a and b, where $\operatorname{lcm}(a,b)$ denotes the least common multiple of a, b.

2.5.3 Show that $\gcd(a,b)\operatorname{lcm}(a,b) = ab$.

2.5.4 Generalise the proof of the prime divisor lemma to show that if $r|ab$ and $\gcd(a,r) = 1$ then $r|b$.

2.6 Congruences

As mentioned in Section 2.3, the fact that $\mathbb{Q}$ is closed under division is very useful, but it does not help us understand the divisibility properties of integers. $\mathbb{Q}$ provides a ratio m/n of any integers m, n with $n \neq 0$, and hence gives no indication whether n actually divides m. What we need is a system containing a more faithful reflection of integer division. Such a system emerges from the concept of *congruence.*

We say that $a, b \in \mathbb{Z}$ are *congruent mod* $n \in \mathbb{Z}$ if n divides $a - b$, and we express this relation by the equation-like expression

$$a \equiv b \pmod{n}$$

called a *congruence mod* n. The word "mod" is short for "modulo," and the number n is sometimes called the *modulus.* The relation $\equiv$ of congruence indeed has the same fundamental properties as equality, namely

$$\begin{aligned} a &\equiv a \pmod{n} && \textit{(reflexive)} \\ a \equiv b \pmod{n} &\Rightarrow b \equiv a \pmod{n} && \textit{(symmetric)} \\ a \equiv b \pmod{n} \text{ and } b \equiv c \pmod{n} &\Rightarrow a \equiv c \pmod{n} && \textit{(transitive)} \end{aligned}$$

as one may easily check. For example, transitivity follows because $n|(b-a)$ and $n|(c-b) \Rightarrow n|(c-b)+(b-a)$, that is, $n|(c-a)$. The difference between $\equiv$ and $=$ is that numbers a, b congruent mod n are not necessarily identical but instead they have a common property – that of leaving the same remainder on division by n.

Thus the congruence notation offers the possibility of computing facts about divisibility in the same way that we manipulate equations. This possibility is realised by the following result, which justifies addition and multiplication of congruences.

Theorem. *If* $a_1 \equiv b_1 \pmod{n}$ *and* $a_2 \equiv b_2 \pmod{n}$ *then*
(i) $a_1 + a_2 \equiv b_1 + b_2 \pmod{n}$
(ii) $a_1 a_2 \equiv b_1 b_2 \pmod{n}$.

Proof. (i) This is straightforward from the definition.

$$\begin{aligned} a_1 \equiv b_1 \pmod{n} &\Rightarrow n|(a_1 - b_1) \\ a_2 \equiv b_2 \pmod{n} &\Rightarrow n|(a_2 - b_2) \\ &\Rightarrow n|(a_1 - b_1) + (a_2 - b_2) \\ &\Rightarrow n|(a_1 + a_2) - (b_1 + b_2) \\ &\Rightarrow a_1 + a_2 \equiv b_1 + b_2 \pmod{n}. \end{aligned}$$

(ii) It is easiest to begin with the special case

$$a \equiv b \pmod{n} \Rightarrow ac \equiv bc \pmod{n} \text{ and } ca \equiv cb \pmod{n}.$$

This holds because

$$\begin{aligned} a \equiv b \pmod{n} &\Rightarrow n|(a-b) \\ &\Rightarrow n|c(a-b) \\ &\Rightarrow n|(ca-cb) \text{ and } n|(ac-bc) \\ &\Rightarrow ca \equiv cb \pmod{n} \text{ and } ac \equiv bc \pmod{n}. \end{aligned}$$

Now to prove the general case we argue as follows:

$$\begin{aligned} a_1 \equiv b_1 \pmod{n} &\Rightarrow a_1a_2 \equiv b_1a_2 \pmod{n} \text{ by the special case } c = a_2 \\ a_2 \equiv b_2 \pmod{n} &\Rightarrow b_1a_2 \equiv b_1b_2 \pmod{n} \text{ by the special case } c = b_1 \\ &\Rightarrow a_1a_2 \equiv b_1b_2 \pmod{n} \text{ by transitivity.} \end{aligned}$$

□

To illustrate the kind of information obtainable by manipulation of congruences, we use congruences mod 9 to explain the ancient rule of "casting out nines": a number (written as usual in base 10 notation) is divisible by 9 ⇔ the sum of its digits is divisible by 9.

If $a_m \ldots a_1a_0$ is the base 10 notation for a number n, then

$$n = 10^m a_m + \cdots + 10a_1 + a_0.$$

Now since $10 \equiv 1 \pmod 9$, we get $10^k \equiv 1^k \equiv 1 \pmod 9$ by multiplication of congruences, and hence, by addition of congruences as well,

$$n = 10^m a_m + \cdots + 10a_1 + a_0 \equiv a_m + \cdots + a_1 + a_0 \pmod 9.$$

In particular,

$$n \equiv 0 \pmod 9 \Leftrightarrow a_m + \cdots + a_1 + a_0 \equiv 0 \pmod 9,$$

that is, 9 divides n ⇔ 9 divides the sum of the digits of n.

Exercises

2.6.1 Use congruences mod 3 to show that

$$3 \text{ divides } n \Leftrightarrow 3 \text{ divides the sum of the digits of } n.$$

2.6.2 Use congruences mod 11 to show that

$$11 \text{ divides } n = a_m \ldots a_1a_0 \Leftrightarrow 11 \text{ divides } (-1)^m a_m + \cdots - a_1 + a_0.$$

2.6.3 Show that it is valid to subtract congruences with the same modulus.

2.6.4 Find an example with $ac \equiv bc \pmod{n}$, with $c \not\equiv 0 \pmod{n}$, but with $a \not\equiv b \pmod{n}$ (that is, division of congruences is *not* always valid).

2.6.5 Use congruences mod 4 to show that $a^2 + b^2 = c^2$ cannot hold for a, b odd and c even.

2.7 Rings and Fields of Congruence Classes

A helpful notation for the set of integers congruent to a mod n is

$$n\mathbb{Z} + a = \{nk + a : k \in \mathbb{Z}\},$$

which we call the *congruence class of a mod n*. Two numbers a, b are congruent mod n if and only if they have the same congruence class, that is,

$$a \equiv b \pmod{n} \Leftrightarrow n\mathbb{Z} + a = n\mathbb{Z} + b. \qquad (*)$$

Thus instead of working with the congruence relation between numbers we can work with the more familiar equality relation between congruence classes. What makes the latter option particularly attractive is that we can define sum and product of congruence classes, and hence treat them as ordinary algebraic objects.

The sum of the class of a_1 and the class of a_2 is defined to be the class of $a_1 + a_2$, and their product is the class of $a_1 a_2$:

$$(n\mathbb{Z} + a_1) + (n\mathbb{Z} + a_2) = n\mathbb{Z} + a_1 + a_2$$
$$(n\mathbb{Z} + a_1)(n\mathbb{Z} + a_2) = n\mathbb{Z} + a_1 a_2.$$

These definitions are meaningful; that is, they are independent of the representatives a_1, a_2 of $n\mathbb{Z} + a_1, n\mathbb{Z} + a_2$, thanks to Theorem 2.6 on the addition and multiplication of congruences. For example, to show that the sum is meaningful we have to show that

$$n\mathbb{Z} + a_1 = n\mathbb{Z} + b_1 \text{ and } n\mathbb{Z} + a_2 = n\mathbb{Z} + b_2 \quad \Rightarrow \quad n\mathbb{Z} + a_1 + a_2 = n\mathbb{Z} + b_1 + b_2,$$

and this is so, because it is equivalent to addition of congruences

$$a_1 \equiv b_1 \pmod{n} \text{ and } a_2 \equiv b_2 \pmod{n} \quad \Rightarrow \quad a_1 + a_2 \equiv b_1 + b_1 \pmod{n}$$

by the equivalence (*). We similarly show that the definition of product is meaningful by appeal to multiplication of congruences.

The sum and product of congruence classes inherit ring properties from the sum and product of integers. For example, the sum of congruence classes is commutative because

$$(n\mathbb{Z} + a_1) + (n\mathbb{Z} + a_2) = n\mathbb{Z} + a_1 + a_2 = n\mathbb{Z} + a_2 + a_1 = (n\mathbb{Z} + a_2) + (n\mathbb{Z} + a_1).$$

The other ring properties are verified similarly, bearing in mind that the zero congruence class is $n\mathbb{Z} + 0 = n\mathbb{Z}$, and the unit congruence class is $n\mathbb{Z} + 1$.

In fact, more is true. We actually have:

Theorem. *The set of congruence classes mod n is a ring under the sum and product of congruence classes. This ring is a field if n is prime.*

Proof. The ring properties all follow from the ring properties of $\mathbb{Z}$, in the same way that commutativity of sum was shown above.

Also, a congruence class $n\mathbb{Z}+a$ has a multiplicative inverse $\Leftrightarrow \gcd(a,n)=1$. The argument goes as follows:

$$\begin{aligned}
& n\mathbb{Z}+a \ \text{ has a multiplicative inverse} \\
\Leftrightarrow\ & (n\mathbb{Z}+a)(n\mathbb{Z}+b) = n\mathbb{Z}+1 \ \text{ for some } b \in \mathbb{Z} \\
\Leftrightarrow\ & n\mathbb{Z}+ab = n\mathbb{Z}+1 \ \text{ for some } b \in \mathbb{Z} \\
\Leftrightarrow\ & ab \equiv 1 \pmod{n} \ \text{ for some } b \in \mathbb{Z} \\
\Leftrightarrow\ & ab-1 = kn \ \text{ for some } b,k \in \mathbb{Z} \\
\Leftrightarrow\ & ab-kn = 1 \ \text{ for some } b,k \in \mathbb{Z} \\
\Leftrightarrow\ & \gcd(a,n) = 1
\end{aligned}$$

(using Theorem 2.4 that $\gcd(a,n) = ab - kn$ for some $b,k \in \mathbb{Z}$ for ($\Leftarrow$), and the fact that a common divisor of a,n also divides $ab-kn$ for ($\Rightarrow$)). Now if n is prime, $\gcd(a,n)=1$ for each a not a multiple of n, that is, for each nonzero $n\mathbb{Z}+a$. Hence each nonzero congruence class has a multiplicative inverse. □

When $\gcd(a,n)=1$ we also say that a and n are *relatively prime*.

The ring of congruence classes mod n, that is, the n-element set $\{n\mathbb{Z}, n\mathbb{Z}+1,\ldots,n\mathbb{Z}+n-1\}$ under the sum and product operations, is often written $\mathbb{Z}/n\mathbb{Z}$. One may also view it as the result of imposing the relation "$n=0$" on $\mathbb{Z}$. The class $n\mathbb{Z}+m$ consists of all numbers that "become equal" to m when n is set equal to 0. $\mathbb{Z}/n\mathbb{Z}$ is also called the *quotient* of $\mathbb{Z}$ by $n\mathbb{Z}$, hence the use of the quotient symbol /. We shall see many other examples of quotient constructions in algebra from Chapter 4 onwards.

The second part of the theorem says that $\mathbb{Z}/p\mathbb{Z}$ is a field when p is prime. Is the converse true? The exercises below show that it is.

Exercises

2.7.1 Show that $\mathbb{Z}/4\mathbb{Z}$ contains nonzero elements whose product is zero (zero divisors).

2.7.2 Show that $\mathbb{Z}/n\mathbb{Z}$ contains zero divisors, and hence is not a field, whenever n is not prime.

2.8* The Theorems of Fermat and Euler

The fact that $\mathbb{Z}/p\mathbb{Z}$ is a field when p is prime has remarkable consequences, justifying the claim at the beginning of Section 2.6 that congruences give a better understanding of integer division. One such consequence is known as *Fermat's little theorem* (to distinguish it from the more famous "Fermat's last theorem").

Fermat's Little Theorem. *If p is prime and* $\gcd(a,p)=1$ *then*

$$a^{p-1} \equiv 1 \ (\textit{mod}\ p).$$

Proof. Consider the congruence classes mod p of $a, 2a, \ldots, (p-1)a$. We shall show that these are distinct and nonzero, and hence they are *all* the $p-1$ nonzero congruence classes mod p. The reason is that $\gcd(a,p)=1$, so a has a multiplicative inverse mod p, by Section 2.7. If we multiply the classes of $a, 2a, \ldots, (p-1)a$ by the inverse of a, we get the distinct nonzero classes of $1, 2, \ldots, p-1$. Hence the classes of $a, 2a, \ldots, (p-1)a$ are also distinct and nonzero.

Thus the congruence classes mod p of $a, 2a, \ldots, (p-1)a$ are the same (though possibly in a different order) as the classes of $1, 2, \ldots, p-1$. It follows that

$$a \times 2a \times \cdots \times (p-1)a \equiv 1 \times 2 \times \cdots \times (p-1) \ (\text{mod}\ p).$$

Now since $1, 2, \ldots, p-1$ are relatively prime to p they all have multiplicative inverses mod p. Multiplying both sides of the congruence by these inverses, we can cancel $1, 2, \ldots, p-1$, obtaining

$$a^{p-1} \equiv 1 \ (\text{mod}\ p). \qquad \square$$

This proof is essentially a modernisation of a proof given by Euler [1761]. Fermat [1640′] only stated the theorem but it seems likely that he found a proof using the binomial theorem, perhaps like the one suggested in the exercises that follow. Euler did not use the language of congruences, which was introduced by Gauss [1801], but he did use the concept of a multiplicative inverse. He also noticed how this concept could be used to generalise the theorem to congruences mod n.

As above, we need $\gcd(a,n)=1$ so that a has a multiplicative inverse mod n. We cannot use all the classes of $1, 2, \ldots, n-1$ mod n, only those that have inverses mod n. These are the classes of the m such that $\gcd(m,n)=1$, and the number of them is denoted by $\phi(n)$, called the *Euler phi function.*

Euler's Theorem. *If* $\gcd(a,n)=1$ *then*

$$a^{\phi(n)} \equiv 1 \ (\textit{mod}\ n).$$

Proof. If we multiply the classes of the m such that $\gcd(m,n)=1$ by a, we get $\phi(n)$ distinct nonzero classes mod n, by the same argument as in Fermat's little theorem. Thus the classes represented by the am, where $\gcd(m,n)=1$, are the same as the classes represented by the m. It follows that

$$\text{product of the } am \equiv \text{ product of the } m \ (\text{mod}\ n),$$

and cancellation of the numbers m from both sides – which we can do since they were chosen to be invertible – yields

$$a^{\phi(n)} \equiv 1 \ (\text{mod}\ n). \qquad \square$$

The relationship between Fermat's little theorem and Euler's theorem is echoed in many developments we shall see later (particularly in the sections marked *). It is typical of the distinction between the "prime case" and the "general case." We shall also see that the field properties of $\mathbb{Z}/p\mathbb{Z}$, and indeed Fermat's little theorem, have important implications for polynomial division (Sections 4.7, 4.8*, 4.9*).

Remark. Fermat's little theorem offers an interesting way to prove that a particular number n is not prime without finding any divisors of n. Namely, find a number a relatively prime to n such that $a^{n-1} \not\equiv 1 \pmod{n}$. Such a number a is called a "witness" to the fact that n is not prime. Finding a witness is not just an interesting possibility – it is also practical, because for large numbers n it is actually faster to compute a^{n-1}, mod n, than to find divisors of n. The secret is to form the powers $a^2, a^4, a^8, a^{16}, \ldots$ by repeated squaring, and then build the exponent $n-1$ by adding suitable powers of 2. For example, to build a^{100} one forms the product $a^{64}a^{32}a^4$. Also, of course, all multiplications are done mod n, so that intermediate results are never much larger than n.

The other fact contributing to the general success of this method is that the number 2 is usually a witness; that is, $2^{n-1} \not\equiv 1 \pmod{n}$ for most nonprime n. The first counterexample is $n = 341$ (see Exercise 2.8.4).

Exercises

2.8.1 Show that p divides the binomial coefficient $\binom{p}{k} = \frac{p!}{k!(p-k)!}$ when $k \neq 0, p$ and p is prime.

2.8.2 Deduce from Exercise 2.8.1 and the binomial theorem that, for p prime,

$$(a+1)^p \equiv a^p + 1 \pmod{p}$$

2.8.3 Deduce by induction on a that

$$a^p \equiv a \pmod{p}$$

for any natural number a, and hence that

$$a^{p-1} \equiv 1 \pmod{p}$$

when $\gcd(a, p) = 1$.

2.8.4 Show that $2^{340} \equiv 1 \pmod{341}$, and hence that the converse to Fermat's little theorem is false.

2.8.5 Is 3 a witness that 341 is nonprime?

2.8.6 Find $\phi(341)$, and check that $2^{\phi(341)} \equiv 3^{\phi(341)} \equiv 1 \pmod{341}$.

2.9* Fractions and the Euler Phi Function

The values of $\phi(n)$ fluctuate wildly and are generally difficult to compute. Nevertheless, ϕ also has some pleasantly regular behaviour which is easily interpreted (and proved) in terms of fractions. There is a natural connection between the two because $\phi(n)$ is the number of *reduced* fractions n'/n where $1 \leq n' \leq n$; that is, the fractions with $\gcd(n', n) = 1$ and hence in "lowest terms."

Theorem. *If m and n are positive integers with $\gcd(m, n) = 1$ then*
(i) $\sum_{d|n} \phi(d) = n$,
(ii) $\phi(mn) = \phi(m)\phi(n)$.

Proof. (i) Consider the n fractions n'/n for $1 \leq n' \leq n$. When n'/n is reduced to lowest terms, the result is a fraction d'/d for some divisor d of n and some $d' \leq d$ which is relatively prime to d. For fixed d, the total number of such fractions d'/d is $\phi(d)$, by definition of ϕ. It is also clear that each fraction d'/d, where $d|n$, results from reducing some n'/n.

Thus to prove that $n = \sum_{d|n} \phi(d)$ it remains to show that distinct reduced fractions d'/d and e'/e (originating from divisors d and e of n, though this doesn't matter) represent distinct rational numbers.

Of course, this is a familiar property of fractions, but it can be proved from first principles as follows. Suppose $d'/d = e'/e$ and hence

$$d'e = de'.$$

This implies $d'|de'$, and hence $d'|e'$ since $\gcd(d', d) = 1$. Similarly $e'|d'$ and hence $d' = e'$. But then we also have $d = e$.

(ii) Consider the $\phi(m)\phi(n)$ sums $\frac{m'}{m} + \frac{n'}{n}$ where $1 \leq m' \leq m$ with $\gcd(m', m) = 1$ and $1 \leq n' \leq n$ with $\gcd(n', n) = 1$. Now

$$\frac{m'}{m} + \frac{n'}{n} = \frac{nm' + mn'}{mn}$$

and the numerator $nm' + mn'$ is relatively prime to mn. Suppose on the contrary that a prime p divides both mn and $nm' + mn'$. Since $p|mn$ and $\gcd(m, n) = 1$ we have $p|m$ or $p|n$. If $p|m$ then $p|mn'$, hence $p|nm' = (nm' + mn') - mn'$, hence $p|m'$ since $p \nmid n$ by the assumption that $\gcd(m, n) = 1$. This contradicts the assumption that m'/m is reduced. There is a similar contradiction if $p|n$.

Thus $\frac{nm'+mn'}{mn}$ is also reduced. Can we get all the $\phi(mn)$ numerators r of reduced fractions with denominator mn in this way? The answer is yes. To find m', n' such that

$$r = nm' + mn',$$

first find m'', n'' such that

$$1 = nm'' + mn'',$$

using the Euclidean algorithm and the fact that $\gcd(m, n) = 1$ (compare with Section 2.4). Then let

$$m' = rm'', \quad n' = rn''.$$

Notice that no prime p divides both m' and m, otherwise it also divides r, contrary to the assumption that r/mn is reduced. Thus m'/m is reduced, and similarly, so is n'/n.

The m', n' we have just found are not necessarily in the ranges $1 \leq m' \leq m$, $1 \leq n' \leq n$ originally considered. However, if $m^* \equiv m' \pmod{m}$ and $n^* \equiv n' \pmod{n}$ are chosen in the required ranges the resulting numerator

$$\begin{aligned} r^* &= nm^* + mn^* \\ &= n(m' + km) + m(n' + ln) \quad \text{for some } k, l \in \mathbb{Z} \\ &\equiv nm' + mn' \pmod{mn}. \end{aligned}$$

Thus we do obtain the $\phi(mn)$ different congruence classes of numerators r from the $\phi(m)\phi(n)$ congruence classes of pairs $(m',\ n')$, which is enough to show $\phi(mn) = \phi(m)\phi(n)$. □

The relation $\phi(mn) = \phi(m)\phi(n)$ when $\gcd(m, n) = 1$ was discovered by Euler [1761] and has since been proved in many different ways. The idea of interpreting it in terms of fractions seems to be due to Kronecker; see Kronecker [1901], p.125. In his book (a set of lecture notes edited by Hensel after Kronecker's death), there is another proof of (ii) which uses (i) and induction (see Kronecker [1901], p.245, or the exercises below). In Chapter 6 we shall give a more abstract, but simpler, proof which also throws light on an ancient result known as the Chinese remainder theorem. A relationship between ϕ and the Chinese remainder theorem was first observed by Gauss [1801], article 38. In article 39 of the same work Gauss also discovered part (i) of the theorem above.

Exercises

2.9.1 Using the fact that $m = \sum_{d|m} \phi(d), n = \sum_{e|n} \phi(e)$, show that

$$mn = \phi(m)\phi(n) + \phi(m) \sum_{e|n, e<n} \phi(e) + \phi(n) \sum_{d|m, d<m} \phi(d) + \sum_{d|m, d<m, e|n, e<n} \phi(d)\phi(e).$$

2.9.2 Assuming $\gcd(m, n) = 1$, show that

$$mn = \sum_{f|mn} \phi(f) = \sum_{d|m, e|n} \phi(de).$$

2.9.3 Use Exercise 2.9.2 and induction on $de < mn$ to show

$$mn = \phi(mn) + \phi(m) \sum_{e|n, e<n} \phi(e) + \phi(n) \sum_{d|m, d<m} \phi(d) + \sum_{d|m, d<m, e|n, e<n} \phi(d)\phi(e),$$

and hence deduce from Exercise 2.9.1 that

$$\phi(mn) = \phi(m)\phi(n) \text{ when } \gcd(m, n) = 1.$$

2.10 Discussion

Euclid's *Elements* is not only a textbook of geometry, but also the first textbook of number theory. Book VII introduces the concepts of divisors and prime numbers, and proves most of their basic properties. Proposition 2 proves the correctness of a simple form of the Euclidean algorithm for finding gcd – repeatedly subtracting the smaller number from the larger instead of dividing by it. Since any common divisor of a, b is also a divisor of $a - b$, it is clear that repeated subtraction gives the gcd just as well as repeated division (though generally more slowly). The advantage of the "division" form of the algorithm is that it works in certain domains where division with remainder cannot always be achieved by repeated subtraction, for example, in rings of polynomials (see Chapter 4).

Despite its early start, number theory did not become a mature discipline until around 1800, with the publication of the first systematic books on the subject: Legendre's *Essai sur la Théorie des Nombres* [1798] and Gauss's *Disquisitiones Arithmeticae* [1801]. Indeed, up to the 18th century most of the important contributions to number theory had been made by two mathematicians: Diophantus (around 250 AD) and Fermat (1601-1665). The surviving writings of Diophantus are nowhere near as general and methodical as the *Elements*. On the face of it they contain nothing but special solutions to special equations. However, these equations and their solutions contain the seeds of much of number theory up to the time of Gauss, the story of which is well told in Weil [1984]. Fermat, in particular, became interested in number theory through reading a book of Diophantus' works edited by Bachet [1621]. Fermat himself worked in the manner of Diophantus, occasionally revealing his solutions but not his general methods, so it was left to Euler, Lagrange and Legendre in the 18th century to prove many of the results claimed or conjectured by Fermat. Their works, and particularly the problems they left unsolved, were the main source of inspiration for Gauss.

Gauss's *Disquisitiones* marks a turning point in the development of number theory because of its appreciation of foundations (such as unique prime factorisation) and abstract algebraic structure (such as congruences, inverses mod p). As mentioned above, Gauss was the first to recognise the importance of unique prime factorisation, even though Euclid had come very close to its proof and other mathematicians were surely aware of it when they spoke of "the" prime factors of a number. Likewise, the notion of congruence is implicit in earlier number theory (for example Fermat's little theorem) but Gauss was first to see that the notion was so pervasive as to deserve its own notation. His own motivation, as he explains in Section I of the *Disquisitiones*, was the problem of deciding whether a given polynomial equation $p(x) = 0$ has a rational solution. In many cases a rational solution can be shown to be nonexistent by considering congruences modulo a suitable m.

Gauss gives the example $p(x) = x^3 - 8x + 6$, for which

$$p(x) \equiv 1, 4, 3, 4, 3 \pmod 5, \quad \text{when} \quad x \equiv 0, 1, 2, 3, 4 \pmod 5.$$

It follows that $p(x) \neq 0$ for any integer x, and hence $p(x) \neq 0$ for any rational x also. The explanation of the last step (assumed known to the reader by Gauss) is that we can assume a rational x to be m/n for relatively prime integers m, n. Then if $p(m/n) = 0$ we have

$$(m/n)^3 - 8(m/n) + 6 = 0$$

hence $m^3 = 8mn^2 - 6n^3 = n^2(8m - 6n)$. Since n divides the right-hand side it must divide m^3, hence any prime divisor of n divides m, by the prime divisor property (Section 2.5). This contradicts the assumption that m, n are relatively prime, unless $n = 1$. Thus any rational solution m/n is in fact an integer solution m. The same argument – rational solution implies integer solution – works for any equation with integer coefficients and leading coefficient 1.

The *Disquisitiones* became the bible of the next generation of number theorists, particularly Dirichlet, who kept a copy of it on his desk at all times. Dirichlet's lectures became a classic in their turn when edited by Dedekind as the book *Vorlesungen über Zahlentheorie*. The first edition appeared in 1863 (four years after Dirichlet's death) and the book gradually changed character as Dedekind added appendices in subsequent editions. The final (4th) edition contains as much Dedekind as Dirichlet. As one would expect from his absorption in the *Disquisitiones*, Dirichlet took pains to clarify and extend the ideas of Gauss. In particular, he gave a simpler treatment of the fundamentals, using the Euclidean algorithm. However, Dirichlet also made great strides into territory where Euler and Gauss had taken only the first steps – the theory of *algebraic* numbers.

An algebraic number is one that satisfies an equation

$$a_n x^n + a_{n-1} x^{n-1} + \cdots + a_1 x + a_0 = 0 \qquad (*)$$

with integer coefficients $a_0, a_1, \ldots, a_n$ (or, equivalently, with rational coefficients). Algebraic numbers are not in general rational, as we shall see in Chapter 3, but despite this they have many properties in common with the rationals. In particular, there are fields of algebraic numbers, and each such field has a ring of *algebraic integers*. Algebraic integers happen to be the numbers that satisfy equations of the form (*) with integer coefficients and $a_n = 1$, though this definition hardly enables one to guess how the algebraic integers of a given field will behave. It turns out that in many cases they behave like ordinary integers – admitting unique factorisation into "primes," for example. This makes it possible to enjoy the greater possibilities of factorisation available with algebraic integers and still draw conclusions about ordinary integers. For example, Euler [1770], p.450, was able to show that there are no positive integers x, y, z such that $x^3 + y^3 = z^3$ (an instance of "Fermat's last theorem") with the help of the factorisation

$$x^3 - z^3 = (x - z)\left(x + \frac{1 + \sqrt{-3}}{2} z\right)\left(x + \frac{1 - \sqrt{-3}}{2} z\right).$$

Dirichlet proved some of the fundamental theorems about algebraic integers, but it was Dedekind who isolated the underlying field and ring properties which explain their similarities with ordinary integers. By making the algebraic structures of number theory the object of study, Dedekind paved the way for abstract algebra. His immediate successors, Weber and Hilbert, still spoke of algebraic number theory rather than ring theory, but the next generation, led by Emmy Noether and Emil Artin, left number theory behind. The definitive account of their teaching, van der Waerden's *Moderne Algebra* [1931], was the first of the "groups, rings and fields" books that are now standard.

It was no doubt an advantage to study rings abstractly as long as their origins in algebraic number theory were still well known, as they were to Noether and Artin in the 1920s. Today, however, it is probably necessary to reiterate that ring theory grew out of the attempt to model the theory of algebraic integers on the theory of ordinary integers. It is therefore not surprising that $\mathbb{Z}$ is a good example of a ring, and one may look to $\mathbb{Z}$ for guidance when investigating the properties of other rings. In Chapter 4 we shall see that several key properties of $\mathbb{Z}$ are reproduced in quite a different setting, the ring of polynomials over a field. This is not even a "modern" discovery, but goes back to Stevin [1585].

Dedekind's reflections on the nature of integers did not end with their algebraic properties. He was also the first to recognise the fundamental role of induction in $\mathbb{N}$. Of course induction has been present, in some sense, from the beginning of number theory. Euclid used it unconsciously, for example in proving that any natural number A has a prime divisor (Book VII, 31). He noted that the process of taking divisors of divisors must eventually halt, otherwise

> an infinite series of numbers will measure the number A, each of which is less than the other: which is impossible in numbers. (Heath [1925], vol. 3, p.332.)

Thus he is appealing to version III of induction as we have described it in Section 2.1. The first to use induction in the "base step, induction step" format was Pascal [1654], who proved several propositions about Pascal's triangle in this way. However, even Gauss and Dirichlet used induction only as an occasional proof technique, without recognising its special character.

Dedekind arrived at a deeper understanding by asking "Was sind und was sollen die Zahlen?" (the German title of his book [1888], which can be translated as "What are numbers and what are they for?"). He came to the conclusion that the essence of the natural number concept is the process of closure under the successor operation, which entails the inductive property of $\mathbb{N}$. He also realised that this property makes it possible to define $+$ and $\times$, so that all of number theory really depends on induction (Dedekind [1888], Theorem 126). This radical rethinking of the nature of number was possible only with the help of the set concept (observe the definition of "closure" in the first paragraph of Section 2.1). In fact many of Dedekind's contributions to mathematics stem from his introduction of sets as mathematical objects. For example, it was his idea to work with congruence *classes*, as algebraic objects, rather than with the congruence *relation* on $\mathbb{Z}$ (Dedekind [1857]). He also used sets to give an elegant definition of real numbers, as we shall see in Chapter 3.

Exercises

2.10.1 Show that the polynomial equation $8x^3 - 6x = 1$ arising from trisection of the angle $\pi/3$ (Section 1.5) has a rational solution only if $y^3 - 3y = 1$ has a rational solution. Show that the latter equation has no rational solution by showing $y^3 - 3y \not\equiv 1 \pmod 2$ for any integer y.

2.10.2 Show that the equation $a_n x^n + \cdots + a_1 x + a_0 = 0$ has a rational solution $x = r/s$ (in lowest terms) only if $r|a_0$ and $s|a_n$.

2.10.3 Use Exercise 2.10.2 to give another proof that $8x^3 - 6x = 1$ has no rational solution.

3 Numbers in General

3.1 Irrational Numbers

The rational numbers are beautifully suited to arithmetic, being closed under $+, -, \times, \div$, but they are inadequate for geometry. Their inadequacy is exposed by the Pythagorean theorem, according to which the diagonal x of the unit square (Figure 3.1.1) satisfies $x^2 = 1^2 + 1^2 = 2$. We can see that no rational number x satisfies this equation as follows.

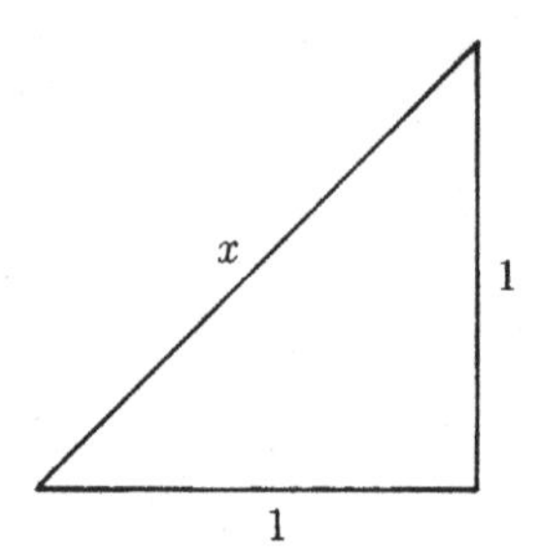

Fig. 3.1.1. Diagonal of the unit square

To satisfy $x^2 = 2$ by $x = m/n$, where $m, n \in \mathbb{Z}$, we need to satisfy

$$m^2/n^2 = 2$$

and hence

$$m^2 = 2n^2.$$

Now it follows from unique prime factorisation (Section 2.5) that the prime factorisation of m^2 contains an even number of 2s (twice the number of 2s in m), and similarly so does the prime factorisation of n^2. But then the prime factorisation of $2n^2$ contains an odd number of 2s, and hence (by unique prime factorisation again) $2n^2$ cannot equal m^2.

Thus the number $x = \sqrt{2}$ is irrational. Or perhaps we should say: if the number $\sqrt{2}$ exists it is irrational. The question of existence will be taken up in Section 3.2.

Similar arguments show that $\sqrt{3}, \sqrt{5}, \sqrt{6}, \ldots$ are irrational. In fact $\sqrt[n]{d}$ is irrational for any integer d which is not an nth power. In particular $\sqrt[3]{2}$ is irrational because $m^3/n^3 = 2$ implies $m^3 = 2n^3$, which is impossible because the number of 2s in the prime factorisation of m^3 is $3\times$something whereas the number of 2s in the prime factorisation of $2n^3$ is $1 + 3\times$something.

A more elaborate argument, still dependent on unique prime factorisation, shows that there is no rational solution to the equation $x^3 - 3x - 1 = 0$ arising from trisection of the angle $\pi/3$ (Section 1.5 and Exercise 2.10.1). It goes as follows.

Suppose for the sake of contradiction that $x^3 - 3x - 1 = 0$ is satisfied by $x = m/n$ where $m, n \in \mathbb{Z}$ and (without loss of generality) that m and n have no common divisor. Substituting $x = m/n$ in the equation we get

$$\frac{m^3}{n^3} - 3\frac{m}{n} - 1 = 0$$

or

$$m^3 - 3mn^2 - n^3 = 0$$

or

$$m^3 = n^2(3m + n).$$

The last equation shows that any prime divisor of n divides the right-hand side, hence it divides m^3, hence it divides m (by the prime divisor lemma, Section 2.5). Since m, n have no common divisor by hypothesis, we must have $n = 1$, and so the last equation simplifies to

$$m^3 = 3m + 1$$

or

$$m(m^2 - 3) = 1.$$

But the product of two integers $m, m^2 - 3$ cannot be 1 unless both are ± 1, which is impossible.

This contradiction shows that there is no rational solution.

Exercises

3.1.1 Prove that $\sqrt{3}$ is irrational.

3.1.2 Prove that if $\sqrt{d} = m/n$ with $m, n \in \mathbb{Z}$ then each prime factor in d occurs to an even power, hence d is a square.

3.1.3 Similarly prove that if $\sqrt[n]{d}$ is rational then d is an nth power.

3.1.4 Is $\log_{10} 2$ rational? Investigate conditions under which $\log_a b$ is rational.

3.1.5 Find two irrational numbers whose sum is rational.

3.1.6 Show that if $e = 1 + \frac{1}{1!} + \frac{1}{2!} + \cdots + \frac{1}{n!} + \cdots$ is rational then $n!e$ is an integer for some n.

3.1.7 Express $n!e$ as an integer plus $\frac{1}{n+1} + \frac{1}{(n+1)(n+2)} + \cdots$, and show that the latter series has sum less than 1.

3.1.8 Deduce from Exercises 3.1.6 and 3.1.7 that e is irrational.

3.2 Existence and Meaning of Irrational Numbers

One possible reaction to the discovery of irrational lengths in geometry is to say that lengths are *not* numbers, but something else. It is then possible to maintain that all numbers are rational, but one is obliged to develop a separate arithmetic of lengths. This was actually done by the Greeks, and they successfully developed an arithmetic of lengths by using comparison with rational lengths.

Since rational multiples of the unit length can approximate any length arbitrarily closely, an irrational length λ is completely determined by the rational lengths $r < \lambda$. For example, length $\sqrt{2}$ is determined by the rational lengths r such that $r^2 < 2$. If μ is another irrational length, determined by the rational lengths $s < \mu$, then $\lambda + \mu$ is determined by the rational lengths $r + s$, and $\lambda\mu$ is determined by the rational lengths rs (we are assuming all lengths are positive here, for simplicity). This is the essence of the arithmetic of lengths developed by Eudoxus around 350 BC and called the "theory of proportions" (see Euclid's *Elements*, Book V).

In 1858 Dedekind noticed that irrationals could be realised without the help of geometry by using *sets* of rational numbers. Since the set of positive rationals r with $r^2 < 2$ determines $\sqrt{2}$, for example, one may as well say it *is* $\sqrt{2}$, and save the trouble of realising $\sqrt{2}$ by a line. Similarly, one can define $\sqrt{3}$, $\sqrt[3]{2}$ etc. to be certain sets of rationals. It is not always easy to state the condition for membership in the set, but any irrational corresponds to a "gap" in $\mathbb{Q}$, and hence to a partition of $\mathbb{Q}$ into a *left set* L of rationals less than the gap, and a *right set* R of rationals greater than the gap. Observing this, Dedekind [1872] defined an *irrational number* λ to be a partition of $\mathbb{Q}$ into non-empty sets L_λ, R_λ such that

(i) each member of L_λ is less than all members of R_λ,
(ii) L_λ has no greatest member and R_λ has no least member,

thus neatly capturing all irrational numbers.

Of course, the set L_λ alone determines λ, as does R_λ alone, and it is actually useful to view λ in this "one-sided" way (see Sections 3.3 and 3.5). The inclusion of both L_λ, R_λ simply makes the definition a little easier to state (see Exercise 3.2.2), and perhaps helps to emphasise the nature of λ as the "gap" between L_λ and R_λ.

Exercises

3.2.1 Use the formula $\frac{\pi}{4} = 1 - \frac{1}{3} + \frac{1}{5} - \frac{1}{7} + \cdots$ to obtain a condition for a rational r to belong to the lower set L for π.

3.2.2 Find conditions equivalent to (i) and (ii) above, but mentioning only the left set L_λ.

3.3 The Real Numbers

Dedekind's definition of irrational numbers is probably as natural and simple as possible, but it does make irrationals look very different from rationals, and we want to view them as the same type of object, namely *real numbers.* This difficulty is easily overcome by viewing each $r \in \mathbb{Q}$ as itself a partition of $\mathbb{Q}$, with left set $L_r = \{q \in \mathbb{Q} : q < r\}$ and right set $R_r = \{q \in \mathbb{Q} : q \geq r\}$. Thus we define a *real number* λ to be a partition of $\mathbb{Q}$ into non-empty sets L_λ, R_λ such that

(i) each member of L_λ is less than all members of R_λ,
(ii) L_λ has no greatest member.

(Alternatively, one can drop (ii) altogether and allow *two* representations of each rational λ.) When the real number λ is viewed as the partition (L_λ, R_λ) it is called a *Dedekind cut.* The set of real numbers is denoted by $\mathbb{R}$.

As Dedekind pointed out, [1872], p.2, when real numbers are defined in this way they behave like points on a line – they have a "left-to-right" order without gaps. The ordering of real numbers $\lambda = (L_\lambda, R_\lambda)$, $\mu = (L_\mu, R_\mu)$ is defined by

$$\lambda \leq \mu \Leftrightarrow L_\lambda \subseteq L_\mu$$

(which is consistent with the ordering of $\mathbb{Q}$ because if $q, r \in \mathbb{Q}$ we have $q \leq r \Leftrightarrow L_q \subseteq L_r$). It follows that

$$\lambda < \mu \Leftrightarrow L_\lambda \subseteq L_\mu \text{ and } L_\lambda \neq L_\mu.$$

Thus if $\lambda < \mu$ there is a rational $q \in L_\mu$ which is not in L_λ, that is, $\lambda < q < \mu$. The informal reason there are no gaps in $\mathbb{R}$ is that any such gap would correspond to a gap in $\mathbb{Q}$, and all such gaps have already been filled by real numbers. The rigorous development of this idea goes as follows.

Theorem. *If $\mathbb{R}$ is partitioned into sets L, R such that any member of L is less than all members of R then either L has a maximum member or R has a minimum member.*

Proof. Suppose $\mathbb{R}$ is partitioned into sets L, R as in the hypothesis of the theorem. This partition of $\mathbb{R}$ induces a cut (L_μ, R_μ) in $\mathbb{Q}$ where

$$L_\mu = \{l \in \mathbb{Q} : l < \text{ some } \lambda \in L\},$$
$$R_\mu = \{r \in \mathbb{Q} : r \geq \text{ some } \rho \in R\}.$$

If the real number $\mu = (L_\mu, R_\mu)$ is less than some $\nu \in L$ we have $\mu < l < \nu$ for some $l \in \mathbb{Q}$, by the definition of $\leq$, hence $l \in L_\mu$ by definition of L_μ and we have the contradiction $l \leq \mu$. Similarly, μ cannot be greater than any $\sigma \in R$.

Thus μ is either the greatest member of L or the least member of R. □

The absence of gaps in $\mathbb{R}$ demonstrates the complete success of Dedekind's definition of real numbers. Not only has $\mathbb{R}$ been defined without appeal to the line, it has all the properties required for it to serve as a *definition* of the line.

Dedekind called the absence of gaps in $\mathbb{R}$ its "continuity," and the property is nowadays called *completeness.*

An equivalent formulation of the completeness of $\mathbb{R}$ is that *every bounded set* $S \subseteq \mathbb{R}$ *has a least upper bound,* that is, a number μ such that $\lambda \in S \Rightarrow \lambda \leq \mu$, and $\kappa < \mu \Rightarrow \kappa < \lambda$ for some $\lambda \in S$. The μ in the proof above is just the least upper bound of L, and similar reasoning shows $L_\mu = \{l \in \mathbb{Q} : l < \text{ some } \lambda \in S\}$ to be the least upper bound of any bounded set $S \subseteq \mathbb{R}$.

Exercises

3.3.1 Explain how each decimal expansion (finite or infinite) determines a real number.

3.3.2 Does this help to explain why $0.999\ldots = 1$?

3.4 Arithmetic and Rational Functions on $\mathbb{R}$

$\mathbb{R}$ inherits the rational operations $+, -, \times, \div$ from $\mathbb{Q}$ in a natural way. For example, if $\lambda = (L_\lambda, R_\lambda)$ and $\mu = (L_\mu, R_\mu)$ then $\lambda + \mu$ is defined by the cut (L, R) where

$$L = \{l + m : l \in L_\lambda, m \in L_\mu\}.$$

If $\lambda, \mu \in \mathbb{Q}$ this agrees with the usual $+$ on $\mathbb{Q}$, because

$$\begin{aligned} \lambda &= \text{ least member of } R_\lambda, \\ \mu &= \text{ least member of } R_\mu, \\ \lambda + \mu &= \text{ least member of } R. \end{aligned}$$

To define multiplication it is easiest to begin with $\lambda, \mu > 0$ and to represent them by cuts $(L_\lambda^+, R_\lambda^+)$, (L_μ^+, R_μ^+) in the *positive* rationals $\mathbb{Q}^+ = \{r \in \mathbb{Q} : r \geq 0\}$. Then $\lambda\mu$ is defined by the cut (L^+, R^+) in $\mathbb{Q}^+$ where

$$L^+ = \{lm : l \in L_\lambda^+, m \in L_\mu^+\}.$$

We can then construct the corresponding cut (L, R) in $\mathbb{Q}$ by expanding L^+ to L with the negative rationals. The definition of multiplication can finally be extended to all reals with the help of the following definition of $-\lambda$. If $\lambda = (L_\lambda, R_\lambda)$ then $-\lambda = (-R_\lambda, -L_\lambda)$, where

$$-S = \{-s : s \in S\} \quad \text{for any } S \subseteq \mathbb{Q}.$$

With these definitions it is easy to check that $\mathbb{R}$ inherits the field properties from $\mathbb{Q}$.

This is only to be expected. What we really want to know is how $\mathbb{R}$ behaves under the *irrational* operations, such as $\sqrt{\ }$, that motivated its construction in the first place. Dedekind remarked that the arithmetic of cuts made possible the first rigorous proof that $\sqrt{2}\sqrt{3} = \sqrt{6}$ (Dedekind [1872], p.22). Indeed, it makes possible the first rigorous proof of the existence of nth roots. The best approach to these results seems to be through the theory of *continuous functions*, a theory which is also fundamental to the algebra of complex numbers (see Section 3.7), hence we shall pause at this point to review the basic properties of continuous functions on $\mathbb{R}$.

A function $f : \mathbb{R} \to \mathbb{R}$ is called *continuous at* $\alpha \in \mathbb{R}$ if, for any $\epsilon > 0$, there is a $\delta > 0$ such that

$$|x - \alpha| < \delta \Rightarrow |f(x) - f(\alpha)| < \epsilon. \qquad (*)$$

We can express this more intuitively by saying that $f(x)$ becomes "arbitrarily close" to $f(\alpha)$ when x is "sufficiently close" to α. The simplest examples of functions continuous at α are constant functions $f(x) = c$ and the identity function $f(x) = x$, which obviously satisfy the condition (*) at each $\alpha \in \mathbb{R}$.

We shall call a function *continuous* if it is continuous at each point of its domain. Thus the identity and constant functions are continuous. The same is true of any function composed from them using $+, -, \times, \div$, that is, any *rational function*, by the following theorem.

Theorem. *If f and g are continuous then so are $f + g, f - g, fg$ and f/g.*

Proof. To prove the continuity of $f + g$ we have to find, for given $\epsilon > 0$, a $\delta > 0$ such that

$$|x - \alpha| \Rightarrow |f(x) + g(x) - f(\alpha) - g(\alpha)| < \epsilon.$$

Since

$$\begin{aligned} |f(x) + g(x) - f(\alpha) - g(\alpha)| &= |f(x) - f(\alpha) + g(x) - g(\alpha)| \\ &\leq |f(x) - f(\alpha)| + |g(x) - g(\alpha)| \end{aligned}$$

it suffices to find a $\delta > 0$ such that

$$|x - \alpha| < \delta \Rightarrow |f(x) - f(\alpha)| < \epsilon/2 \text{ and } |g(x) - g(\alpha)| < \epsilon/2.$$

By the continuity of f we can make $|f(x) - f(\alpha)| < \epsilon/2$ for $|x - \alpha| < \delta_1$, say. And by the continuity of g we can make $|g(x) - g(\alpha)| < \epsilon/2$ for $|x - \alpha| < \delta_2$, say. Hence it suffices to take $\delta = \min(\delta_1, \delta_2)$. The continuity of $f - g$ is proved similarly.

To prove the continuity of fg we have to make $|f(x)g(x) - f(\alpha)g(\alpha)| < \epsilon$. The trick is to use

$$\begin{aligned} |f(x)g(x) - f(\alpha)g(\alpha)| &= |f(x)g(x) - f(x)g(\alpha) + f(x)g(\alpha) - f(\alpha)g(\alpha)| \\ &\leq |f(x)||g(x) - g(\alpha)| + |g(\alpha)||f(x) - f(\alpha)|. \end{aligned}$$

We can assume $|f(x)| \leq$ some constant A, say $|f(\alpha)| + 1$, by confining x to some distance δ_A from α. Then it suffices to make

$$|g(x) - g(\alpha)| < \epsilon/2A \text{ and } |f(x) - f(\alpha)| < \epsilon/2|g(\alpha)|$$

which we can do as above by appealing to the continuity of f and g.

To prove the continuity of f/g, having proved the continuity of products, it suffices to prove the continuity of $1/g$. To do this we have to make $\frac{1}{g(x)} - \frac{1}{g(\alpha)} < \epsilon$. Since

$$\left|\frac{1}{g(x)} - \frac{1}{g(\alpha)}\right| = \left|\frac{g(x)-g(\alpha)}{g(x)g(\alpha)}\right| = \frac{1}{|g(x)||g(\alpha)|}|g(x)-g(\alpha)|,$$

and since $g(\alpha) \neq 0$ if α is in the domain of $1/g$, we can make $|g(x)| \geq$ some constant $B > 0$ by confining x to some distance δ_B from α. Then it suffices to make $|g(x) - g(\alpha)| \leq B|g(\alpha)|\epsilon$, which we can do by the continuity of g. □

Exercise

3.4.1 Prove $\sqrt{2}\sqrt{3} = \sqrt{6}$ using the definition of product of Dedekind cuts.

3.5 Continuity and Completeness

As mentioned in Section 3.3, Dedekind used the word "continuity" to describe the absence of gaps in $\mathbb{R}$, not to describe a property of functions, as we now do. But where does the concept of a continuous function come from, after all? It comes from trying to capture the notion of a function *whose graph has no gaps.* It is clear from the case of a strictly increasing function (Figure 3.5.1) that a function whose graph has no gaps has to be continuous in the ϵ–δ sense. But the proof that the graph of a continuous function has no gaps depends on the "continuity," that is, the completeness, of $\mathbb{R}$.

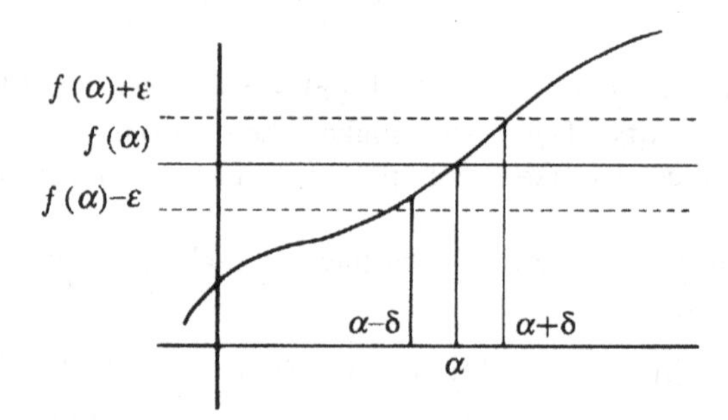

Fig. 3.5.1. Graph of a continuous function

A precise formulation of this "no gap" property of continuous functions is in the following theorem. Notice how convenient it is to express completeness as the existence of least upper bounds.

Intermediate Value Theorem. *If f is continuous at all points of a closed interval $[\alpha, \beta] = \{x : \alpha \leq x \leq \beta\}$ and if $f(\alpha) < 0$ and $f(\beta) > 0$ then there is a $\gamma \in [\alpha, \beta]$ such that $f(\gamma) = 0$.*

Proof. Let $S = \{\mu \in [\alpha, \beta] : f < 0 \text{ on } [\alpha, \mu]\}$. Then S is bounded above by β, and hence it has a least upper bound γ by Section 3.3. We shall show that $f(\gamma) = 0$ by showing that the possibilities $f(\gamma) > 0$ and $f(\gamma) < 0$ lead to contradiction.

If $f(\gamma) = \nu > 0$ then by continuity of f we can find a $\delta > 0$ such that $f(x) > 0$ for $\gamma - \delta < x < \gamma + \delta$ (take $\epsilon = \nu/2$). In particular $f(\mu) > 0$ for some $\mu < \gamma$, contrary to the definition of γ (that $f < 0$ on $[\alpha, \mu]$ for all $\mu < \gamma$).

If $f(\gamma) = \varphi < 0$ we can similarly find a $\delta > 0$ such that $f(x) < 0$ for $\gamma - \delta < x < \gamma + \delta$. Since $f < 0$ on $[\alpha, \mu]$ for any $\mu < \gamma$ by definition of γ it follows that we also have $f < 0$ on $[\alpha, \mu]$ for some $\mu > \gamma$, which is also contrary to the definition of γ. □

Corollary (existence of nth roots). *For any $\lambda > 0$ and integer $n > 0$ there is a γ such that $\gamma^n = \lambda$.*

Proof. Consider the function $f(x) = x^n - \lambda$ on the closed interval $[0, 1 + \lambda]$. It follows from Theorem 3.4 that $f(x)$ is continuous for all x in $[0, 1 + \lambda]$. We clearly have $f(0) = -\lambda < 0$, and $f(1 + \lambda) = (1 + \lambda)^n - \lambda > 0$ (consider the two cases $\lambda < 1$ and $\lambda \geq 1$ separately).

Hence by the intermediate value theorem there is a γ such that $f(\gamma) = \gamma^n - \lambda = 0$, that is, $\gamma^n = \lambda$. □

Another important consequence of completeness is the following theorem, which we shall generalise to the plane in Section 3.7.

Extreme Value Theorem. *If f is continuous on $[\alpha, \beta]$ then f attains maximum and minimum values on $[\alpha, \beta]$.*

Proof. We first prove that f is bounded on $[\alpha, \beta]$.

Suppose on the contrary that f is unbounded on $[\alpha, \beta]$. Then it is unbounded on at least one of the intervals $[\alpha, \frac{\alpha+\beta}{2}]$ (lower half of $[\alpha, \beta]$), $[\frac{\alpha+\beta}{2}, \beta]$ (upper half of $[\alpha, \beta]$). Let $[\alpha_1, \beta_1]$ be the lower of the two halves on which f is unbounded, and argue similarly on $[\alpha_1, \beta_1]$, obtaining a half $[\alpha_2, \beta_2]$ of $[\alpha_1, \beta_1]$ on which f is unbounded, and so on.

In this way we obtain a sequence of intervals $[\alpha_1, \beta_1] \supset [\alpha_2, \beta_2] \supset [\alpha_3, \beta_3] \supset \cdots$ with a single common point λ. In fact λ is the least upper bound of $\{\alpha_1, \alpha_2, \ldots\}$ because

$$\alpha_1 \leq \alpha_2 \leq \cdots \leq \lambda \leq \cdots \leq \beta_2 \leq \beta_1$$

and $\beta_i - \alpha_i$ becomes arbitrarily small as i increases. Now, by definition of λ, any of its neighborhoods $\lambda - \delta < x < \lambda + \delta$ contains an interval $[\alpha_i, \beta_i]$ on which f is unbounded. But this is absurd, since f is continuous and therefore we can make $|f(x)| \leq |f(\gamma)| + \epsilon$ by suitable choice of δ.

This contradiction proves that $f(x)$ is bounded and hence the set $\{f(x) : x \in [\alpha, \beta]\}$ has a least upper bound μ. Now for the coup de grâce. If $f(x)$ does *not* take the value μ then $\frac{1}{\mu - f(x)}$ is continuous on $[\alpha, \beta]$ by Theorem 3.4 and unbounded since $f(x)$ takes values arbitrarily close to μ. This contradicts the boundedness of continuous functions just proved! Hence f attains a maximum value, μ.

Similarly, f attains a minimum value. □

Exercises

3.5.1 Use the intermediate value theorem to show that any odd-degree polynomial equation with real coefficients has a real root.

3.5.2 Show that any continuous f which maps $[\alpha, \beta]$ into itself has a *fixed point*, that is, a γ such that $f(\gamma) = \gamma$.

3.6 Complex Numbers

Complex numbers are objects of the form $\alpha + i\beta$ where $\alpha, \beta \in \mathbb{R}$ and $i^2 = -1$. There is of course no $i \in \mathbb{R}$ whose square is -1, so this is a further extension of the number system. The set $\{\alpha + i\beta : \alpha, \beta \in \mathbb{R}\}$ of complex numbers is denoted by $\mathbb{C}$. There is a unique extension of $+, -, \times, \div$ to $\mathbb{C}$ satisfying the field properties, since if these properties hold we necessarily have

$$\begin{aligned}(\alpha_1 + i\beta_1) + (\alpha_2 + i\beta_2) &= (\alpha_1 + \alpha_2) + i(\beta_1 + \beta_2),\\ (\alpha_1 + i\beta_1)(\alpha_2 + i\beta_2) &= (\alpha_1\alpha_2 - \beta_1\beta_2) + i(\alpha_1\beta_2 + a_2\beta_1)\end{aligned}$$

(using the fact that $i^2 = -1$). Conversely, it can be checked that the sum and product defined by these equations have the field properties.

A nice application of the field properties of $\mathbb{C}$ is to show that a product of sums of two squares, $(\alpha_1^2 + \beta_1^2)(\alpha_2^2 + \beta_2^2)$, is itself a sum of two squares. Viewing $\alpha_1^2 + \beta_1^2$ and $\alpha_2^2 + \beta_2^2$ as *differences* of two squares in $\mathbb{C}$, we get the factorisations

$$\begin{aligned}\alpha_1^2 + \beta_1^2 &= \alpha_1^2 - i^2\beta_1^2 = (\alpha_1 - i\beta_1)(\alpha_1 + i\beta_1),\\ \alpha_2^2 + \beta_2^2 &= \alpha_2^2 - i^2\beta_2^2 = (\alpha_1 - i\beta_2)(\alpha_2 + i\beta_2),\end{aligned}$$

whence

$$\begin{aligned}(\alpha_1^2 + \beta_1^2)(\alpha_2^2 + \beta_2^2) &= [(\alpha_i - i\beta_1)(\alpha_2 - i\beta_2)][(\alpha_1 + i\beta_1)(\alpha_2 + i\beta_2)]\\ &= [\alpha_1\alpha_2 - \beta_1\beta_2 - i(\alpha_1\beta_2 + \alpha_2\beta_1)][\alpha_1\alpha_2 - \beta_1\beta_2 + i(\alpha_1\beta_2 + \alpha_2\beta_1)]\\ &= (\alpha_1\alpha_2 - \beta_1\beta_2)^2 + (\alpha_1\beta_2 + \alpha_2\beta_1)^2. \qquad (*)\end{aligned}$$

Once discovered, of course, this identity between real numbers can be checked by multiplying out both sides. However, it does look a lot more natural in $\mathbb{C}$.

Since each $\alpha + i\beta \in \mathbb{C}$ is uniquely determined by the pair (α, β) of reals, we can interpret $\mathbb{C}$ as the set of *points* (α, β), that is, as a *plane*. Just as the real number α captures the intuitive idea of a point on a line, the pair (α, β) captures the idea of the point in the plane with cartesian coordinates α, β.

We shall now sketch how the arithmetic of $\mathbb{C}$ captures the geometry of the plane. Naturally, the basic definitions are motivated by background knowledge of geometry. Nevertheless, it is a surprise to see how geometrically effective $+$ and $\times$ become when combined with $i = \sqrt{-1}$.

The distance of a point $z = \alpha + i\beta$ from $O = (0,0)$ is captured by its *absolute value*

$$|z| = |\alpha + i\beta| = \sqrt{\alpha^2 + \beta^2}.$$

This is motivated by the Pythagorean theorem (see Figure 3.6.1). More generally, $|z_2 - z_1|$ captures the distance between z_1 and z_2, and there is an algebraic proof of the *triangle inequality*:

$$|z_3 - z_1| \leq |z_3 - z_2| + |z_2 - z_1|.$$

Another important property of absolute value is

$$|z_1 z_2| = |z_1||z_2|,$$

which comes from the identity (*) when $z_1 = \alpha_1 + i\beta_1$ and $z_2 = \alpha_2 + i\beta_2$.

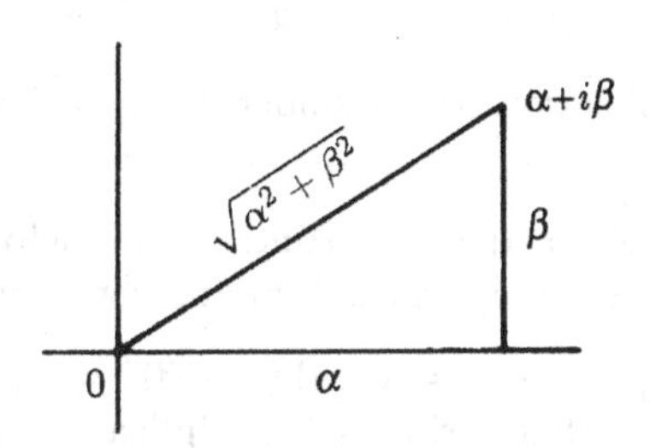

Fig. 3.6.1. Distance from the origin

With this natural definition of distance, addition and multiplication turn out to be geometrically significant mappings of the plane. Addition of a constant $\alpha + i\beta$ is simply the *translation* of the plane sending every point (x, y) a constant distance, to the point $(x + \alpha, y + \beta)$. To understand multiplication by $\alpha + i\beta$, we first rewrite it in its *polar form*

$$\alpha + i\beta = |\alpha + i\beta| \left(\frac{\alpha}{|\alpha + i\beta|} + \frac{i\beta}{|\alpha + i\beta|} \right),$$

which is the product of the real number $|\alpha + i\beta| = \sqrt{\alpha^2 + \beta^2}$ and the complex number $\frac{\alpha+i\beta}{|\alpha+i\beta|} = \frac{\alpha}{\sqrt{\alpha^2+\beta^2}} + \frac{i\beta}{\sqrt{\alpha^2+\beta^2}}$ of absolute value 1. Multiplication by $|\alpha+i\beta|$ is the mapping called *dilatation* of the plane by factor $|\alpha + i\beta|$. It magnifies all distances by the constant factor $|\alpha + i\beta|$. Multiplication by $\frac{\alpha+i\beta}{|\alpha+i\beta|}$ is called a *rotation* of the plane about O. Thus multiplication in general is the composite of a dilatation with a rotation about O.

The "rotation" terminology can be reconciled with intuitive geometry by observing that any complex number of absolute value 1 can be written in the form $\cos\theta + i\sin\theta$, and that

$$(x+iy)(\cos\theta + i\sin\theta) = (x\cos\theta - y\sin\theta) + i(x\sin\theta + y\cos\theta)$$

is indeed the point to which $x+iy$ is sent by a counter-clockwise rotation through θ. However, now that we are constructing geometry from $\mathbb{R}$, rather than the other way round, we have to eschew geometrically defined functions like cos and sin. A more direct justification for the term "rotation" is that multiplication by any complex number of absolute value 1 fixes the origin and preserves all distances (see Exercises).

Exercises

3.6.1 If x', y' are defined by $x' + iy' = (x+iy)(\lambda + i\mu)$ show that

$$(x_2' - x_1')^2 + (y_2' - y_1')^2 = (\lambda x_2 - \lambda x_1 - \mu y_2 + \mu y_1)^2 + (\mu x_2 - \mu x_1 + \lambda y_2 - \lambda y_1)^2$$

and that this

$$= (x_2 - x_2)^2 + (y_2 - y_1)^2 \quad \text{when} \quad \lambda^2 + \mu^2 = 1.$$

Conclude that multiplication by a number $\lambda + i\mu$ of absolute value 1 preserves distances.

3.6.2 Prove the triangle inequality. (Hint: it may help to reduce to the special case $z_1 = 0$, $z_2 \in \mathbb{R}$ by suitable translation and rotation.)

3.6.3 Suppose $|\alpha+i\beta| = |\gamma+i\delta| = 1$ and $(\alpha+i\beta)^3 = \gamma+i\delta$, so that multiplication by $\alpha + i\beta$ is a rotation that "trisects" multiplication by $\gamma + i\delta$. Show that the relation between α and γ is $4\alpha^3 - 3\alpha = \gamma$ (compare with Section 1.5).

3.6.4 Show that any composite of translations, rotations and dilatations of the plane is a function of the form $f(z) = az + b$. Is the converse also true?

3.7 Regular Polygons

The relationship between rotation and the multiplication of complex numbers throws new light on the nature of the regular n-gon. It enables us to justify the claim, made in Section 1.5, that constructing the regular n-gon is equivalent to constructing the solutions of

$$z^{n-1} + \cdots + z + 1 = 0. \qquad (*)$$

Indeed, it will become clear that the *existence* of the regular n-gon is equivalent to the existence of solutions to (*).

We take the plane to be $\mathbb{C}$, choose the unit of length so that the n-gon is inscribed in the unit circle $\{z : |z| = 1\}$, and choose the real axis so that one vertex is at $z = 1$. The other vertices result from 1 by a sequence of rotations, that is, multiplications by some ζ with $|\zeta| = 1$, and hence they form a sequence of complex numbers $1, \zeta, \zeta^2, \ldots, \zeta^{n-1}$ where $\zeta^n = 1$. Of course we have in mind that $\zeta = \cos\frac{2\pi}{n} + i\sin\frac{2\pi}{n}$ (Figure 3.7.1), but in accordance with our program of constructing geometry from numbers we avoid using cos and sin. The only thing we need to know about ζ is that $\zeta^n = 1$, reflecting the fact that the product of n rotations (multiplications by ζ) returns each vertex to its starting point, and hence equals multiplication by 1. Bearing in mind that

$$\zeta^n - 1 = (\zeta - 1)(\zeta^{n-1} + \cdots + \zeta + 1)$$

and $\zeta \neq 1$, we see that $\zeta^{n-1} + \cdots + \zeta + 1 = 0$. Thus if the regular n-gon exists in $\mathbb{C}$, there is a complex number ζ satisfying (*).

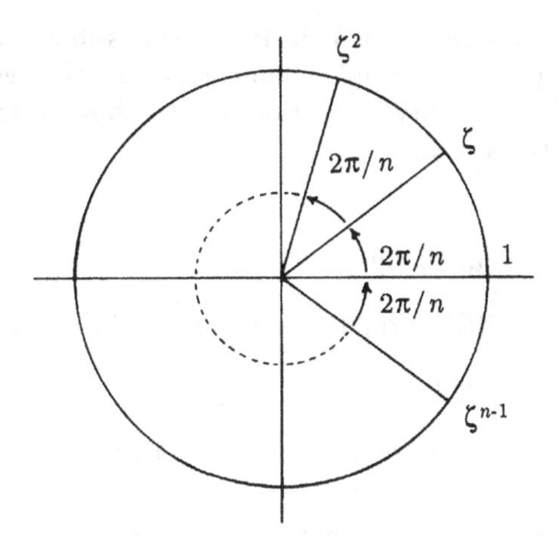

Fig. 3.7.1. Vertices of the regular n-gon

Conversely, does a solution ζ of (*) yield a regular n-gon? Well, not just any solution will do. For example, the solution $z = i$ of

$$z^7 + z^6 + \cdots + z + 1 = 0$$

gives a regular 4-gon (with vertices $i, i^2, i^3, i^4 = 1$), not a regular 8-gon. In general, if $n = md$ a solution ζ of (*) may satisfy not just $\zeta^n = 1$ but also $\zeta^d = 1$, in which case ζ gives at most a d-gon, not an n-gon. To get an n-gon we need what is called a *primitive* n^{th} *root of unity*, a $\zeta = \zeta_n$ such that $\zeta_n^n = 1$ but $\zeta_n^d \neq 1$ for $d < n$. In this case $\zeta_n, \zeta_n^2, \ldots, \zeta_n^{n-1}, 1$ are distinct and may be taken as the vertices of the n-gon.

When n is a prime p, then any solution ζ of (*) is a primitive p^{th} root of unity, because if $\zeta^d = 1$ and $d < n$ we have $\zeta^{dm} = (\zeta^d)^m = 1$ for any integer m, and by taking m to be the multiplicative inverse of d, mod p (see Section

2.7), we get $1 = \zeta^{dm} = \zeta^1$, which is a contradiction. Thus when n is prime the existence of the regular n-gon follows from the existence of *any* solution to (*). In this case we can also prove (see Section 4.7) that $z^{p-1} + \cdots + z + 1$ has no non-trivial factorisation with rational coefficients. We call $z^{p-1} + \cdots + z + 1$ the pth *cyclotomic polynomial* ("cyclotomic" is from the Greek word meaning "circle-dividing") and denote it by $\Phi_p(z)$.

When n is not prime the left-hand side of (*) factorises into polynomials with integer coefficients, one of which, $\Phi_n(z)$, is satisfied only by primitive n^{th} roots of unity. We shall not go further into this here, though Exercises 3.7.3 and 3.7.4 give some clues.

At any rate, assuming the existence of a $\Phi_n(z)$ satisfied only by primitive n^{th} roots of unity, we can say that the existence of the regular n-gon follows from the existence of solutions to the equation

$$\Phi_n(z) = 0,$$

called the nth *cyclotomic equation.* Thus it is now possible to base the regular n-gon on algebraic properties of complex numbers, rather than geometric intuition. We shall realise this possibility in Section 3.8, by showing that *all* polynomial equations have solutions in $\mathbb{C}$.

Exercises

3.7.1 Defining integers α_n, β_n by

$$(3 + 4i)^n = \alpha_n + i\beta_n \quad \text{for} \quad n = 1, 2, 3, \ldots,$$

show by induction on n that

$$\alpha_n \equiv 3 \pmod 5 \quad \text{and} \quad \beta_n \equiv 4 \pmod 5.$$

3.7.2 Deduce from Exercise 3.7.1 that $(\frac{3}{5} + \frac{4}{5}i)^n \neq 1$, that is, the point $\frac{3}{5} + \frac{4}{5}i$ on the unit circle is not a root of unity.

3.7.3 Use the fact that

$$z^{n-1} + \cdots + z + 1 = (z^n - 1)/(z - 1)$$

to find a factorisation of $z^{n-1} + \cdots + z + 1$ when n is not prime.

3.7.4 Defining $\Phi_n(z)$ inductively by

$$\Phi_1(z) = z - 1, \quad \Phi_n(z) = (z^n - 1)/\prod_{d|n} \Phi_d(n) \quad \text{for} \quad n > 1,$$

show that $\Phi_4(z) = z^2 + 1$ and $\Phi_6(z) = z^2 - z + 1$.

3.7.5 If $\gcd(m, n) = 1$ and the regular m-gon and the regular n-gon are constructible, show that the regular mn-gon is constructible.

3.8 The Fundamental Theorem of Algebra

The statement that every polynomial equation with coefficients in $\mathbb{C}$ has a solution in $\mathbb{C}$ is called the fundamental theorem of algebra. All known proofs, however, depend on the non-algebraic concept of continuity. This is not surprising, and it is certainly no excuse to omit the theorem from a book on algebra! We shall now give a proof which uses a minimum of algebra, based on a generalisation to $\mathbb{C}$ of the extreme value theorem of Section 3.5. (The minimum of continuity required to prove the theorem seems to be the intermediate value theorem, used in Section 3.5 to prove solvability of the special equation $x^n = \lambda \in \mathbb{R}$.)

The first step is the following lemma, which depends only on the geometric properties of $\mathbb{C}$.

d'Alembert's Lemma. *If $p(z)$ is a nonconstant polynomial and $p(z_0) \neq 0$ then any neighborhood of z_0 contains a z_1 such that $|p(z_1)| < |p(z_0)|$.*

Proof. We let $z_1 = z_0 + \Delta z$, where Δz is a complex number to be determined so that $|p(z_1)| < |p(z_0)|$, that is, so that $p(z_1)$ is nearer to the origin than $p(z_0)$. If

$$p(z) = a_n z^n + a_{n-1}z^{n-1} + \cdots + a_1 z + a_0$$

then

$$p(z_0 + \Delta z) = a_n z_0^n + a_{n-1}z_0^{n-1} + \cdots + a_1 z_0 + a_0 + A_1\Delta z + A_2(\Delta z)^2 + \cdots$$

for some constants $A_1, A_2, \ldots$ (not all zero, since p is nonconstant)

$$= p(z_0) + A\Delta z + \epsilon,$$

where $A = A_i(\Delta z)^{i-1}$ contains the first nonzero A_i and $|\epsilon|$ is small compared with $|A\Delta z|$ when $|\Delta z|$ is small (because ϵ contains the higher powers of Δz).

It is then clear from Figure 3.8.1 that by choosing the direction of Δz so that $A\Delta z$ is opposite in direction to $p(z_0)$, and choosing $|\Delta z|$ so that $|\epsilon|$ is much smaller than $|A\Delta z|$, we get $p(z_1) = p(z_0) + A\Delta z + \epsilon$ closer to O than $p(z_0)$, that is, $|p(z_1)| < |p(z_0)|$. □

Now to prove the fundamental theorem we consider the real-valued function $|p(z)|$ of z in $\mathbb{C}$. Using the obvious definition of continuity of a function f at $w \in \mathbb{C}$, namely

$$|z - w| < \delta \Rightarrow |f(z) - f(w)| < \epsilon,$$

one sees that the identity and constant functions are continuous. Arguments like those in Section 3.4 prove the continuity of sums and products of continuous functions, hence a polynomial $p(z)$ is continuous and so is $|p(z)|$.

The analogue of the extreme value theorem (Section 3.5) for $\mathbb{C}$, that a continuous function $|f|$ on a closed disc $|z| \leq \mathbb{R}$ attains a maximum and a minimum, has an analogous proof by construction of a sequence of nested squares with a single common point. Figure 3.8.2 gives the idea.

We shall now prove the fundamental theorem by showing that the minimum of $|p(z)|$, over a sufficiently large disc $|z| \leq R$, is zero. (If $|p(z)| = 0$ then, necessarily, $p(z) = 0$.)

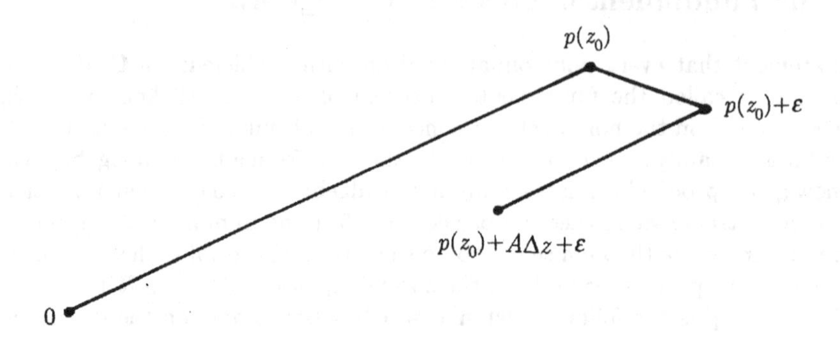

Fig. 3.8.1. A suitable choice of Δz

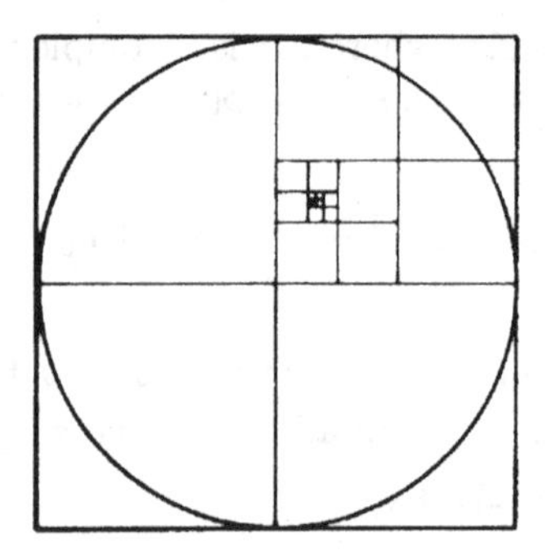

Fig. 3.8.2. Nested squares

Fundamental Theorem of Algebra. *If $p(z)$ is a polynomial in z with coefficients in $\mathbb{C}$ then there is a number $c \in \mathbb{C}$ with $p(c) = 0$.*

Proof. Let $p(z) = a_n z^n + a_{n-1}z^{n-1} + \cdots + a_1 z + a_0$. Then for $|z|$ sufficiently large, $|a_n z^n|$ is large in comparison with $|a_{n-1}z^{n-1} + \cdots + a_1 z + a_0|$, so $|p(z)|$, like $|a_n z^n|$, increases with $|z|$. Suppose this happens for $|z| \geq R$.

Since $|p(z)|$ is continuous, it takes a minimum value over the disc $|z| \leq R$, at $z = c$ say. Now if the minimum $|p(c)| \neq 0$ we get a contradiction by d'Alembert's lemma as follows. If $|c| < R$, so c has a neighborhood inside the disc $|z| \leq R$, then the z_1 in this neighborhood with $|p(z_1)| < |p(c)|$ contradicts the minimality of $|p(c)|$. If $|c| = R$, then the z_1 with $|p(z_1)| < |p(c)|$ contradicts minimality in the disc if $|z_1| \leq R$, and it contradicts the increase of $|p(z)|$ if $|z_1| > R$.

Thus $|p(c)| = 0$ and hence $p(c) = 0$. □

Remark. An important result related to the fundamental theorem, but much more elementary, is the following:

Factor Theorem (Descartes [1637], p.159). *If $p(z)$ is a polynomial and $p(c) = 0$ then $p(z) = (z-c)q(z)$ for some polynomial $q(z)$.*

Proof. For each positive integer m we have the factorisation

$$z^m - c^m = (z-c)(z^{m-1} + z^{m-2}c + \cdots + zc^{m-2} + c^{m-1}),$$

which may be verified by multiplying out the right-hand side. Thus if

$$p(z) = a_n z^n + \cdots + a_1 z + a_0$$

we have

$$\begin{aligned} p(z) - p(c) &= a_n(z^n - c^n) + \cdots + a_1(z-c) \\ &= (z-c)q(z) \end{aligned}$$

for some polynomial $q(z)$. And since $p(c) = 0$ by hypothesis this gives $p(z) = (z-c)q(z)$. □

Corollary. *If $p(z)$ has degree n then the equation $p(z) = 0$ has at most n solutions.*

Proof. Given one solution $z = c$ we have

$$p(z) = (z-c)q(z),$$

in which case the other solutions are the solutions of $q(z) = 0$. Since $q(z)$ has degree $n-1$, the corollary follows by induction on degree. □

Notice that the proof of the factor theorem does not involve the continuity of polynomials, only their algebraic properties. The algebraic setting of the theorem will be explored in the next chapter, where it will be proved again and generalised.

Exercises

3.8.1 Suppose that $p(z)$ has real coefficients. Show
(i) $p(\alpha + i\beta) = 0 \Rightarrow p(\alpha - i\beta) = 0$ when $\alpha, \beta \in \mathbb{R}$.
(ii) $p(z)$ factorises into real linear and quadratic factors.

3.8.2 Find the real quadratic factors of $x^4 + 1$.

3.9 Discussion

The effective use of the set concept in mathematics probably begins with Dedekind's definition of real numbers in 1858. As mentioned in Section 2.9, Dedekind went on to use sets systematically as mathematical objects, giving meaning to previously dubious, unrecognised or ill-defined concepts. Set theory has now become indispensable in this role, but its effectiveness does not end there. It has also contributed a remarkable proof technique – the *diagonal argument* – which gives the simplest known proof that not all real numbers are algebraic. Nonalgebraic, or *transcendental*, numbers were first discovered by Liouville [1844]. He was able to show that the number

$$\sum_{n=1}^{\infty} 10^{n!} = 0.1100010000\ldots$$

is transcendental by proving a theorem on the approximation of algebraic numbers by rationals. A construction of transcendental numbers by set theory was given by Cantor [1874], using nothing but the definition of algebraic numbers. In his paper [1891] he introduced the diagonal argument which makes the proof even simpler.

The definition of algebraic numbers is required only to establish a one-to-one correspondence between them and the natural numbers. One first lists the equations

$$a_n x^n + \cdots + a_1 x + a_0 = 0 \qquad (*)$$

that define algebraic numbers in order of a quantity Cantor called their "height," namely

$$h = |a_n| + \cdots + |a_1| + |a_0| + n.$$

Since $a_0, a_1, \ldots, a_n$ are integers there are only finitely many equations of a given height. Hence, listing them in order of increasing height gives a sequence in which each equation appears at some natural number position. Then if each equation (*) is replaced by its roots (finitely many, by Section 3.8), the result is a sequence in which each algebraic number appears at some natural number position. This gives the desired one-to-one correspondence between the set of algebraic numbers and $\mathbb{N}$.

The diagonal construction is now used to show that, for any infinite list $x_0, x_1, x_2, \ldots$ of real numbers, it is possible to construct a real number x not on the list. The simplest way is to take the decimal expansions of $x_0, x_1, x_2, \ldots$ and make x differ suitably from each x_n in the nth decimal place. For example, one can let the nth decimal place of x be 2 if the nth decimal place of x_n is 1, and let it be 1 otherwise. (This avoids getting a decimal expansion which is ultimately $000\ldots$ or $999\ldots$, and hence possibly equal to some x_n, though differently expressed.) This construction is called "diagonal" because it involves just the digits along the diagonal in the obvious tabulation of the decimal expansions of $x_0, x_1, x_2, \ldots$. If the diagonal digits are underlined the construction of $x \neq x_0, x_1, x_2, \ldots$ might look like this, for example:

$$\begin{aligned} x_0 &= 3.\underline{0}1010\ldots \\ x_1 &= 0.1\underline{1}111\ldots \\ x_2 &= 1.14\underline{1}42\ldots \\ x_3 &= 2.222\underline{2}2\ldots \\ x_4 &= 1.2599\underline{2}\ldots \\ &\;\;\vdots \\ x &= 0.12211\ldots \end{aligned}$$

In particular, when $x_0, x_1, x_2, \ldots$ is a list of all real algebraic numbers, x is a transcendental number.

One could even compute a particular transcendental number by listing the equations (*) according to some rule, computing their real solutions in order and to a sufficient number of decimal places, and then obtaining the successive decimal places of x as above. Admittedly, one would not learn anything interesting about x by doing this – it merely shows that Cantor's proof is not a pure existence proof. If one wants to see a comprehensible transcendental number then one needs to prove something about algebraic numbers, such as Liouville's approximation theorem. Even more work is required to show that the familiar numbers e and π are transcendental. The Lindemann [1882] proof of the transcendence of π, mentioned in Section 1.9, is based on analytic methods developed by Hermite [1873] to prove the transcendence of e. (Of course, none of this is intended to belittle the diagonal construction. It is not only the quickest route to transcendental numbers, it also reveals the extraordinary fact that $\mathbb{R}$ is a strictly larger set than $\mathbb{N}$, in the sense that $\mathbb{N}$ cannot be put in one-to-one correspondence with $\mathbb{R}$. The other constructions of transcendental numbers give no hint of this.)

The question of existence versus construction also arises with the fundamental theorem of algebra. Early attempts to prove the theorem were incomplete, mainly because they failed to reckon with the existential part of the proof – usually an application of the extreme value theorem or the intermediate value theorem. The proof given above, which requires the extreme value theorem, is based on results of d'Alembert [1746] and Argand [1806]. They glossed over the existence of the extreme value. Likewise, proofs by Laplace [1795] and Gauss [1816] require the intermediate value theorem, but gloss over it. Mathematics up to this time had not encountered a situation where existence was not obtained by an explicit construction. The first to recognise the need for the intermediate value theorem was Bolzano [1817], but he was unable to prove it rigorously because he lacked a definition of real numbers.

Only after definitions of $\mathbb{R}$ (such as Dedekind's) became known could Weierstrass [1874] give rigorous proofs of the intermediate value theorem and the extreme value theorem, and thus complete the proof of the fundamental theorem of algebra. As can be seen from Section 3.5, the proofs are quite simple, but *not* constructive. This is unavoidable for general continuous functions. For polynomial functions it is possible to construct the required value by an infinite sequence of approximations, but not simple.

Fortunately these practical difficulties are irrelevant when it comes to studying the behavior of algebraic numbers. We do not need to know the decimal expansion of $\sqrt{2}$, for example, to know that

$$(\sqrt{2})^2 = 2,$$
$$(\sqrt{2})^3 = 2\sqrt{2},$$
$$(\sqrt{2})^3 + \sqrt{2} = 3\sqrt{2}, \quad \text{etc.}$$

All the algebraic properties of $\sqrt{2}$ are consequences of the equation $x^2 = 2$ satisfied by $x = \sqrt{2}$. Similarly, the behavior of any algebraic number α follows from a polynomial equation $p(x) = 0$ satisfied by $x = \alpha$. The fundamental theorem of algebra is relevant only when one wants to know that there is a *number* satisfying a given polynomial equation $p(x) = 0$. The existence and construction of an *abstract* solution of $p(x) = 0$ is almost a triviality, as we shall see in Section 4.5. All we need is a little experience with polynomials, which will be acquired in Sections 4.1 to 4.4.

The contrast between the construction of an abstract algebraic solution of $p(x) = 0$ and the construction of a complex number solution makes it more understandable why the "fundamental theorem of algebra" is given short shrift in most algebra books. It is not so much a theorem about polynomials as a theorem about the complex numbers. Since the complex numbers are defined by the "infinitary" methods of analysis (Dedekind cuts or convergent series), it is not surprising that the fundamental theorem requires such methods for its proof.

There is even some historical justice in this conclusion. The first serious attempts to prove the theorem, beginning with d'Alembert [1746], actually had the aim of solving a problem in calculus – the integration of $1/p(x)$. To do this it suffices to factorise $p(x)$ into linear or quadratic factors; the integration can then be completed by the method of partial fractions. Early versions of the fundamental theorem, such as Gauss [1799], explicitly asserted the factorisation of a real polynomial into real linear or quadratic factors. To see what this has to do with complex roots, see Exercise 3.8.1.

The special case of this integration problem where $p(x) = x^n - 1$ also led to the first result relating the regular n-gon to the equation $x^n - 1 = 0$. Cotes [1722] found a geometric realisation of the factorisation of $x^n - 1$ into real linear and quadratic factors by interpreting the roots of $x^n - 1 = 0$ as points on the unit circle.

Exercise

3.9.1 Find the real quadratic factors of $x^n - 1$ by suitable grouping of the linear factors, obtained by de Moivre's theorem.

4 Polynomials

4.1 Polynomials over a Field

Up until now we have viewed polynomials as functions of a real or complex variable x. This viewpoint is important for the fundamental theorem of algebra (Section 3.8), for example. Quite a different insight is obtained if, as in high-school algebra, we refrain from giving any interpretation to the letter x, and simply study the formal behavior of polynomials under $+, -$ and $\times$. Since

$$\begin{aligned}(a_0 + a_1x + a_2x^2 + \cdots) &+ (b_0 + b_1x + b_2x^2 + \cdots)\\ &= (a_0 + b_0) + (a_1 + b_1)x + (a_2 + b_2)x^2 + \cdots,\\ (a_0 + a_1x + a_2x^2 + \cdots) &- (b_0 + b_1x + b_2x^2 + \cdots)\\ &= (a_0 - b_0) + (a_1 - b_1)x + (a_2 - b_2)x^2 + \cdots\end{aligned}$$

and

$$\begin{aligned}(a_0 + a_1x + a_2x^2 + \cdots) &\times (b_0 + b_1x + b_2x^2 + \cdots)\\ &= a_0b_0 + (a_0b_1 + a_1b_0)x + (a_0b_2 + a_1b_1 + a_2b_0)x^2 + \cdots\end{aligned}$$

it becomes clear that the behavior of polynomials is really the behavior of their sequences of coefficients. In fact, we could *define* a polynomial to be a sequence $(a_0, a_1, a_2, \ldots)$ with only a finite number of nonzero terms, and define $+, -, \times$ on sequences to agree with the equations above:

$$\begin{aligned}(a_0, a_1, a_2, \ldots) + (b_0, b_1, b_2, \ldots) &= (a_0 + b_0, a_1 + b_1, a_2 + b_2, \ldots),\\ (a_0, a_1, a_2, \ldots) - (b_0, b_1, b_2, \ldots) &= (a_0 - b_0, a_1 - b_1, a_2 - b_2, \ldots),\\ (a_0, a_1, a_2, \ldots) \times (b_0, b_1, b_2, \ldots) &= (a_0b_0, a_0b_1 + a_1b_0, a_0b_2 + a_1b_1 + a_2b_0, \ldots).\end{aligned}$$

This is actually convenient for computer algebra, but for humans it seems better to continue using the letter x. To emphasize its purely formal, place-holding role, x is called an *indeterminate*.

At any rate, it is clear from this discussion that the properties of polynomials depend entirely on the set from which their coefficients are chosen. The set of coefficients should be closed under $+, -, \times$ to allow us to form $+, -, \times$ of polynomials, hence it should at least be a ring. The set of polynomials in an indeterminate x with coefficients in a ring R is denoted by $R[x]$. It follows easily from the equations above that $R[x]$ itself is a ring, hence we speak of the *ring* $R[x]$ of polynomials. For example, $\mathbb{Z}[x]$ is the ring of polynomials in x with integer coefficients.

$\mathbb{Z}[x]$ unfortunately lacks some of the properties of familiar rings like $\mathbb{Z}$, notably the property of division with remainder. It turns out that to perform division of polynomials with remainder (long division) one has to be able to divide coefficients. Thus the problem can be rectified by going from $\mathbb{Z}[x]$ to $\mathbb{Q}[x]$, and in general by considering the ring $F[x]$ of polynomials with coefficients in a field F. Mostly we will be concerned with fields $F \subseteq \mathbb{C}$, the so-called *number fields*, but crucial use will also be made (in Sections 4.7 and 4.9*) of polynomials with coefficients in the finite fields $\mathbb{Z}/p\mathbb{Z}$ from Section 2.7.

Exercises

4.1.1 Use long division of polynomials to find quotient and remainder when $x^2 + 1$ is divided by $2x + 1$.

4.1.2 Are there a reasonable "quotient" $q(x)$ and "remainder" $r(x)$ in $\mathbb{Z}[x]$ such that $x^2 + 1 = q(x)(2x + 1) + r(x)$?

4.1.3 Show that $F[x]$ is not a field.

4.2 Divisibility

The definition of divisibility in $F[x]$ is the same as the definition of divisibility in $\mathbb{Z}$, namely, if $f, g \in F[x]$ then

$$g \text{ divides } f \Leftrightarrow f = gh \text{ for some } h \in F[x].$$

In fact, apart from some obvious differences which arise at the outset, the theory of divisibility in $F[x]$ is very similar to the theory of divisibility in $\mathbb{Z}$. We can therefore take advantage of our experience with $\mathbb{Z}$ (Chapter 2) to penetrate quickly to the important theorems (Sections 4.3 and 4.4).

The obvious differences between $\mathbb{Z}$ and $F[x]$ are in the *units* (the elements that divide everything) and the measure of size (in $\mathbb{Z}$, the absolute value). The units in $\mathbb{Z}$ are ± 1, whereas the units in $F[x]$ are all the nonzero elements of F (the "constant" polynomials). From this it is clear that absolute value is not an appropriate measure of size in $F[x]$, but fortunately there is another measure – the *degree* of the polynomial, that is, the exponent of the highest power of x.

The concept of degree is crucial to the following property, which unlocks the whole theory of divisibility in $F[x]$.

Division Property of $F[x]$. *If $f, g \in F[x]$ and $g \neq 0$ then there are $q, r \in F[x]$ with*

$$f = qg + r \text{ and } \operatorname{degree}(r) < \operatorname{degree}(g).$$

Proof. Consider the set of natural numbers $\{\operatorname{degree}(f - qg) : q \in F[x]\}$. By version III of induction (Section 2.1), this set has a least member. Let n be the least member and let $f - qg$ be a polynomial of degree n. Thus

$$f - qg = a_n x^n + \text{ lower degree terms.}$$

Also suppose

$$g = b_m x^m + \text{ lower degree terms.}$$

Now if $n \geq m$ the polynomial $f - qg - \frac{a_n}{b_m} x^{n-m} g = f - q'g$ has lower degree than $f - qg$, contrary to the definition of n.

Thus there is an $r = f - qg \in F[x]$ with $\operatorname{degree}(r) = n < \operatorname{degree}(g) = m$. $\square$

The brief but essential appearance of division in this proof – forming $\frac{a_n}{b_m}$ – is the reason we require F to be a field. The use of induction is essentially a condensation of the long division process for polynomials, which subtracts multiples qg of g from f, with q chosen so as to repeatedly remove the highest degree terms from f.

The division property gives a less computational proof of the factor theorem of Section 3.8.

Factor Theorem. *If $p(x) \in F[x]$ and $p(c) = 0$ for some $c \in F$ then $p(x) = (x - c)q(x)$ for some $q(x) \in F[x]$.*

Proof. By the division property, with $g(x) = x - c$,

$$p(x) = (x - c)q(x) + r(x)$$

for some $r(x) \in F[x]$ with degree$(r) <$ degree$(x - c)$, hence $r(x) =$ constant, $k \in F$. Substituting $x = c$ we get

$$0 = p(c) = r(c) = k,$$

and therefore $p(x) = (x - c)q(x)$. □

An element c such that $p(c) = 0$ is called a *root of the equation* $p(x) = 0$, or simply a *root of* $p(x)$.

Corollary. *If $p(x) \in F[x]$ has degree d then $p(x)$ has at most d roots in F.*

Proof. For each root c there is a factor $x - c$ of $p(x)$, and a polynomial of degree d cannot have more than d such factors. □

The analogue of a prime in $F[x]$ – a polynomial divisible only by constants or by constant multiples of itself – is called an *irreducible* polynomial. We often say irreducible *over* F to emphasize the field of coefficients available for divisors. For example, $x^2 - 2$ is irreducible over $\mathbb{Q}$ but not over $\mathbb{R}$, since

$$x^2 - 2 = (x - \sqrt{2})(x + \sqrt{2}) \quad \text{and} \quad \pm\sqrt{2} \notin \mathbb{Q} \quad \text{but} \quad \pm\sqrt{2} \in \mathbb{R}.$$

With the help of the concept of irreducibility we can generalise the factor theorem above to the case where c does not necessarily belong to F. The generalisation tells us, for example, that $x^2 - 2$ is a factor of any rational $p(x)$ satisfied by $x = \sqrt{2}$.

General Factor Theorem. *If $p(x) \in F[x]$ and $p(c) = 0$, and if $h(x) \in F[x]$ is an irreducible polynomial such that $h(c) = 0$, then $p(x) = h(x)q(x)$ for some $q(x) \in F[x]$.*

Proof. First we show that an irreducible $h(x) \in F[x]$ with root c is unique up to a constant factor. To do this, suppose $h^*(x) \in F[x]$ is a polynomial *of minimal degree* such that $h^*(c) = 0$, and consider the following consequences of the division property.

$$
\begin{aligned}
h(x) &= q^*(x)h^*(x) + r^*(x) \\
&\qquad \text{where } q^*(x), r^*(x) \in F[x] \text{ and } \mathrm{degree}(r^*) < \mathrm{degree}(h^*) \\
&\Rightarrow 0 = h(c) = q^*(c)h^*(c) + r^*(c) \\
&\Rightarrow r^*(c) = 0 \ \text{ since } h^*(c) = 0 \\
&\Rightarrow r^*(x) = 0 \ \text{ since } \mathrm{degree}(r^*) < \mathrm{degree}(h^*) \text{ and } h^* \text{ is of minimal degree} \\
&\Rightarrow q^*(x) = \text{constant}, \quad \text{since } h \text{ is irreducible.}
\end{aligned}
$$

Thus all irreducible $h(x)$ satisfied by c are constant multiples of a fixed $h^*(x)$, and hence of each other.

If we now divide $p(x)$ by $h(x)$ the division property gives $q(x), r(x) \in F[x]$ with

$$p(x) = h(x)q(x) + r(x) \text{ and } \mathrm{degree}(r) < \mathrm{degree}(h),$$

and $r(x) = 0$ by the minimality of $h(x)$ we have just established. Thus

$$p(x) = h(x)q(x) \text{ with } q(x) \in F[x],$$

as required. □

It is obvious that a divisor of a divisor is a divisor, and that dividing by a nontrivial divisor lowers degree, hence it follows as in $\mathbb{Z}$ that any $f \in F[x]$ factorises into irreducibles. Also as in $\mathbb{Z}$, factorisation into irreducibles is not quite unique, but it is unique up to constant factors, as we shall see in Section 4.3.

Exercises

4.2.1 Exhibit infinitely many irreducibles in $F[x]$, none of which is a constant multiple of another.

4.2.2 Call $f \in \mathbb{Q}[x]$ a *minimal polynomial* (over $\mathbb{Q}$) for $\alpha \in \mathbb{C}$ if $f(\alpha) = 0$ and $g(\alpha) \neq 0$ for each $g \in \mathbb{Q}[x]$ of lower degree than f. Use the division property to show that if $h \in \mathbb{Q}[x]$ and $h(\alpha) = 0$ then f divides h.

4.2.3 Deduce from Exercise 4.2.2 that

$$f \in \mathbb{Q}[x] \text{ is minimal for } \alpha \in \mathbb{C} \Leftrightarrow f(\alpha) = 0 \text{ and } f \text{ is irreducible over } \mathbb{Q}.$$

4.2.4 Find a minimal polynomial for $\alpha = 1 + \sqrt{2}$ over $\mathbb{Q}$.

4.2.5 Show that $1 + \sqrt{2} = \sqrt{3 + 2\sqrt{2}}$.

4.2.6 If $\sqrt{r + s\sqrt{t}}$ is the root of an irreducible quartic $f \in \mathbb{Q}[x]$ and $\sqrt{r + s\sqrt{t}} = \sqrt{a} + \sqrt{b}$, where $a, b, r, s, t \in \mathbb{Q}$, show that the set of four elements $\pm\sqrt{r \pm s\sqrt{t}}$ is the same as the set of four elements $\pm\sqrt{a} \pm \sqrt{b}$. (Hint: find rational quartics satisfied by $\sqrt{r + s\sqrt{t}}$ and $\sqrt{a} + \sqrt{b}$ and use uniqueness.)

4.3 Unique Factorisation

Thanks to the division property, the Euclidean algorithm works in $F[x]$ the same way as in $\mathbb{N}$. Beginning with polynomials f, g we divide f by g and let $f_1 = g$, $g_1 =$ remainder when f is divided by g, thus obtaining an f_1 "larger" (in degree) than g_1. At each subsequent stage k the "larger" polynomial f_k is divided by the "smaller" polynomial $g_k = f_{k+1}$, obtaining a remainder g_{k+1} which is "smaller" than f_{k+1}. Since "size" (degree) cannot decrease indefinitely, eventually g_k divides f_k exactly, in which case $g_k = \gcd(f, g)$ by the same argument as for $\mathbb{N}$.

Moreover, since both new polynomials

$$f_{k+1} = g_k,\ g_{k+1} = f_k - q_k g_k \text{ for some } q_k \in F[x],$$

are linear combinations of the old with coefficients in $F[x]$ we can conclude, as in $\mathbb{N}$, that

$$\gcd(f, g) = uf + vg \text{ for some } u, v \in F[x].$$

The analogue of the prime divisor lemma (Section 2.5) is that *if an irreducible* p *divides* fg *then* p *divides* f *or* p *divides* g, and the proof is exactly the same as for $\mathbb{N}$. One only has to bear in mind that if f is not a multiple of p we can say $1 = \gcd(f, p)$. Of course *any* nonzero constant can now be taken as the gcd of f and p, but 1 is certainly allowed.

Finally, this lemma gives unique factorisation into irreducibles much as the prime divisor lemma gives unique factorisation in $\mathbb{N}$. We suppose

$$p_1 \cdots p_r = q_1 \cdots q_s$$

are two factorisations into irreducibles, and conclude that p_1 divides some q_i. It no longer follows that $q_i = p_1$, only that $q_i = c_i p_1$ for some $c_i \in F$, but this is good enough. We can then cancel p_1 from the equation, leaving c_i in place of q_i on the right-hand side. Repetition of the process until all of $p_1, \ldots, p_r$ are cancelled leads to the conclusion that they are the same (up to order) as $c_1 q_1, \ldots, c_s q_s$ for some $c_1, \ldots, c_s \in F$ (in particular, $r = s$). Thus we have:

Theorem. *Factorisation of a polynomial into irreducibles over a field F is unique up to order and constant factors.* □

Exercises

4.3.1 Show that any factorisation $k(x - \alpha_1) \cdots (x - \alpha_n)$ of a polynomial into linear factors is unique (up to the order of factors).

4.3.2 Formulate a unique factorisation theorem for polynomials with real coefficients.

4.4 Congruences

In $F[x]$, as in $\mathbb{Z}$, the meaning of $f \equiv g \pmod{l}$ is that l divides $f-g$. The basic properties of congruence follow as before, but now using the ring properties of $F[x]$. In particular

$$\begin{aligned} f &\equiv f \pmod{l}, \\ f \equiv g \pmod{l} &\Rightarrow g \equiv f \pmod{l}, \\ f \equiv g \pmod{l} \text{ and } g \equiv h \pmod{l} &\Rightarrow f \equiv h \pmod{l}, \\ f_1 \equiv g_1 \pmod{l} \text{ and } f_2 \equiv g_2 \pmod{l} &\Rightarrow f_1 + f_2 \equiv g_1 + g_2 \pmod{l}, \\ f_1 \equiv g_1 \pmod{l} \text{ and } f_2 \equiv g_2 \pmod{l} &\Rightarrow f_1 f_2 \equiv g_1 g_2 \pmod{l}. \end{aligned}$$

The latter two properties imply, as in Section 2.7, that we can define $+$ and $\times$ of *congruence classes of polynomials*. The congruence class of $f \in F[x]$ mod l is

$$lF[x] + f = \{lk + f : k \in F[x]\},$$

and the sum and product of congruence classes are defined by

$$\begin{aligned} (lF[x] + f) + (lF[x] + g) &= lF[x] + (f + g) \\ (lF[x] + f) \times (lF[x] + g) &= lF[x] + (f \times g), \end{aligned}$$

hence they inherit the ring properties from $F[x]$.

Finally, we find as in Section 2.7 that

$$lF[x] + f \text{ has a multiplicative inverse} \Leftrightarrow \gcd(f, l) = 1,$$

using the validity of the Euclidean algorithm in $F[x]$ (Section 4.2) to show that

$$uf + vl = 1 \text{ for some } u, v \in F[x] \Leftrightarrow \gcd(f, l) = 1.$$

The Euclidean algorithm gives the $(\Leftarrow)$ direction, the $(\Rightarrow)$ direction holds because a common divisor of f, l is also a common divisor of $uf + vl$, that is, of 1.

Now if l is an irreducible polynomial p we have $\gcd(f, p) = 1$ for each $f \not\equiv 0 \pmod{p}$, hence we have the following analogy with $\mathbb{Z}$ mod a prime:

Theorem. *The set $F[x]/lF[x]$ of congruence classes mod $l \in F[x]$ is a ring under the sum and product of congruence classes. This ring is a field if l is irreducible.* □

Exercises

4.4.1 Show that each $f \in F[x]$ is congruent mod $l \in F[x]$ to a polynomial of degree less than degree(l).

4.4.2 Show that if $f, g \in F[x]$ have degree less than degree(l) and $f \neq g$ then $f \not\equiv g \pmod{l}$.

4.4.3 Show that $F[x]/lF[x]$ has zero divisors when l is not irreducible.

4.5 The Fields $F(\alpha)$

From now on we shall write a polynomial $p \in F[x]$ as $p(x)$ to avoid confusion with a number p (and similarly for polynomials denoted by other letters).

The fields $F[x]/p(x)F[x]$ for irreducible $p(x) \in F[x]$ are analogous to the fields $\mathbb{Z}/p\mathbb{Z}$ for prime $p \in \mathbb{Z}$, but they are actually more familiar, at least when F is a familiar field such as $\mathbb{Q}$. $F[x]/p(x)F[x]$ is the same as the field $F(\alpha)$ that results from *adjoining a root* α *of* $p(x)$ *to* F, that is, the closure of $F \cup \{\alpha\}$ under $+, -, \times, \div$ (by a nonzero element). In particular, if $F \subseteq \mathbb{C}$ then we can take $\alpha \in \mathbb{C}$ by the fundamental theorem of algebra (Section 3.8), so $F[x]/p(x)F[x]$ has a concrete interpretation $F(\alpha)$ as a field of numbers.

Before studying an example, note the use of brackets. Square brackets are used for the *ring* $F[x]$, that is, the closure of $F \cup \{x\}$ under $+, -, \times$; round brackets (or parentheses) are used for the *field* $F(\alpha)$, that is, the closure of $F \cup \{\alpha\}$ under $+, -, \times, \div$.

Example. $\mathbb{Q}[x]/(x^2-2)\mathbb{Q}[x]$.

We recognise this field as $\mathbb{Q}(\sqrt{2})$ by interpreting x as $\sqrt{2}$. This interpretation gives all polynomials in the congruence class of $f(x)$ the value $f(\sqrt{2})$, since they all differ from $f(x)$ only by multiples of x^2-2, which is zero when $x=\sqrt{2}$. Thus the elements of $\mathbb{Q}[x]/(x^2-2)\mathbb{Q}[x]$ correspond to certain numbers in $\mathbb{Q}(\sqrt{2})$.

In fact, they correspond to numbers of the form $a+b\sqrt{2}$, where $a, b \in \mathbb{Q}$, since each $f(x) \in \mathbb{Q}[x]$ is congruent to an element of the form $a+bx$, namely its remainder on division by x^2-2. And the numbers $a+b\sqrt{2}$, $a'+b'\sqrt{2}$ corresponding to different remainders are also different, by the irrationality of $\sqrt{2}$. (If $a+b\sqrt{2}=a'+b'\sqrt{2}$ we must have $a=a'$ and $b=b'$, because if $b \neq b'$ we have the contradiction $\sqrt{2}=(a-a')/(b'-b) \in \mathbb{Q}$, and if $b=b'$ then $a=a'$.)

Moreover, *every* number in $\mathbb{Q}(\sqrt{2})$ is of the form $a+b\sqrt{2}$ with $a, b \in \mathbb{Q}$. One way to see this is to check that numbers of this form are closed under $+, -, \times, \div$. This is easy for $+, -, \times$ and for $\div$ it follows from the fact that

$$\frac{1}{a+b\sqrt{2}}=\frac{1}{a+b\sqrt{2}} \cdot \frac{a-b\sqrt{2}}{a-b\sqrt{2}}=\frac{a}{a^2-2b^2}-\frac{b}{a^2-2b^2}\sqrt{2}.$$

Thus the correspondence between congruence classes in $\mathbb{Q}[x]/(x^2-2)\mathbb{Q}[x]$ and numbers in $\mathbb{Q}(\sqrt{2})$ is one-to-one. Finally, the correspondence preserves sums and products. The value of a sum $f(x)+g(x)$ is the sum $f(\sqrt{2})+g(\sqrt{2})$ of the values, and value of a product $f(x)g(x)$ is the product $f(\sqrt{2})g(\sqrt{2})$ of the values, hence corresponding elements have the same algebraic behavior. □

A one-to-one correspondence which preserves $+$ and $\times$ is called an *isomorphism* (of fields, in this case). It captures the sense in which $\mathbb{Q}[x]/(x^2-2)\mathbb{Q}[x]$ is the "same" as $\mathbb{Q}(\sqrt{2})$. By generalising the correspondence between polynomial classes and their values we can similarly find an isomorphism expressing the "sameness" of $F[x]/p(x)F[x]$ and $F(\alpha)$ whenever $p(x) \in F[x]$ is irreducible and α is a root of $p(x)$.

Theorem. *If F is a field, $p(x) \in F[x]$ is irreducible over F, and α is a root of $p(x)$, then $F[x]/p(x)F[x]$ is isomorphic to $F(\alpha)$.*

Proof. First consider the evaluation map $\tilde{\sigma} : F[x] \to F(\alpha)$ defined by

$$\tilde{\sigma}(f(x)) = f(\alpha).$$

This map $\tilde{\sigma}$ preserves $+$ and $\times$ and gives the same value $f(\alpha)$ to all members of the congruence class $p(x)F[x] + f(x)$ of $f(x)$. Hence $\tilde{\sigma}$ induces a map $\sigma : F[x]/p(x)F[x] \to F(\alpha)$ defined by

$$\sigma(p(x)F[x] + f(x)) = f(\alpha),$$

which also preserves $+$ and $\times$. It remains to show that σ maps $F[x]/p(x)F[x]$ one-to-one onto $F(\alpha)$.

Now if $p(x)$ has degree n, each congruence class $p(x)F[x]+f(x)$ has a unique member of the form $a_0 + a_1x + \cdots + a_{n-1}x^{n-1}$ with $a_0, \cdots, a_{n-1} \in F$, namely the remainder when $f(x)$ is divided by $p(x)$. The corresponding elements $a_0 + a_1\alpha + \cdots + a_{n-1}\alpha^{n-1}$ are distinct, because if

$$a_0 + a_1\alpha + \cdots + a_{n-1}\alpha^{n-1} = a_0' + a_1'\alpha + \cdots + a_{n-1}'\alpha^{n-1}$$

and some $a_i \neq a_i'$ then $(a_0 - a_0') + \cdots + (a_n - a_n')x^{n-1}$ is a nonzero polynomial satisfied by α, contrary to the minimality of irreducible polynomials found in Section 4.2. Thus σ is one-to-one.

Finally, σ is onto $F(\alpha)$ because each element of $F(\alpha)$ is of the form $a_0 + a_1\alpha+\cdots+a_{n-1}\alpha^{n-1}$. Elements of this form comprise a subset of $F(\alpha)$ obviously closed under $+, -$, and closed under $\times$ because $\alpha^n, \alpha^{n+1}, \ldots$ can be rewritten as combinations of $1, \alpha, \ldots, \alpha^{n-1}$ using the equation $p(\alpha) = 0$. To show closure under $\div$ it suffices to find an inverse of any $a_0 + a_1\alpha + \cdots + a_{n-1}\alpha^{n-1}$. The inverse is necessarily the value $b_0 + b_1\alpha + \cdots + b_{n-1}\alpha^{n-1}$ of the polynomial $b_0 + b_1x + \cdots + b_{n-1}x^{n-1}$ inverse to $a_0 + a_1x + \cdots + a_{n-1}x^{n-1}$ mod $p(x)$, since

$$\begin{aligned}
1 = \tilde{\sigma}(1) &= \tilde{\sigma}((a_0 + a_1x + \cdots + a_{n-1}x^{n-1})(b_0 + b_1x + \cdots + b_{n-1}x^{n-1})) \\
&= \tilde{\sigma}(a_0 + a_1x + \cdots + a_{n-1}x^{n-1})\tilde{\sigma}(b_0 + b_1x + \cdots + b_{n-1}x^{n-1}) \\
&= (a_0 + a_1\alpha + \cdots + a_{n-1}\alpha^{n-1})(b_0 + b_1\alpha + \cdots + b_{n-1}\alpha^{n-1})
\end{aligned}$$

by definition of $\tilde{\sigma}$ and its preservation of $\times$. □

The proof contains the following useful description of $F(\alpha)$ – which on the face of it is a set of rational functions of α – as a set of polynomials in α.

Corollary. *If $p(x) \in F[x]$ is irreducible and of degree n then*

$$F(\alpha) = \{a_0 + a_1\alpha + \ldots + a_{n-1}\alpha^{n-1} : a_0, \ldots, a_{n-1} \in F\}. \qquad \square$$

Remark. If F is a field in which not every polynomial $p(x)$ has a root, we may wonder where to find a root α of $p(x)$. The theorem tells us that, when $p(x)$ is irreducible, we need look no further than the field $F[x]/p(x)F[x]$. In $F[x]/p(x)F[x]$ the congruence class of x satisfies $p(x) = 0$ by definition, and the theorem says that this congruence class has the same algebraic behavior as any concrete root α. It is rather amusing to have a root of $p(x)$ for the asking, so to speak, but there is a small catch – we have to know that $p(x)$ is irreducible, otherwise $F[x]/p(x)F[x]$ is not a field. We turn to the problem of recognising irreducibility in the next two sections.

Exercises

4.5.1 Show that if $x^3 - 2$ is not irreducible over $\mathbb{Q}$ then $x^3 = 2$ has a rational root.

4.5.2 Conclude from Exercise 4.5.1 and Section 3.1 that

$$\mathbb{Q}(\sqrt[3]{2}) = \{a + b\sqrt[3]{2} + c(\sqrt[3]{2})^2 : a, b, c \in \mathbb{Q}\}$$

4.5.3 Show that $x^4 - 4 = 0$ has no rational root but that $x^4 - 4$ is not irreducible over $\mathbb{Q}$.

4.6 Gauss's Lemma

Gauss's lemma answers a natural question about polynomials with *integer* coefficients, that is, polynomials in $\mathbb{Z}[x]$. If

$$f(x) = x^n + a_{n-1}x^{n-1} + \cdots + a_1x + a_0,$$

where $a_0, a_1, \ldots, a_{n-1}$ are integers, and if $f(x)$ has a factorisation

$$f(x) = g(x)h(x)$$

where

$$g(x) = x^r + b_{r-1}x^{r-1} + \cdots + b_1x + b_0, \quad h(x) = x^s + c_{s-1}x^{s-1} + \cdots + c_1x + c_0$$

are in $\mathbb{Q}[x]$, must $b_0, \ldots, b_{r-1}, c_0, \ldots, c_{s-1}$ be integers? (The restriction to polynomials with leading coefficient 1, or *monic* polynomials as they are called, is necessary to avoid trivial counterexamples.)

The answer to this question is yes, as Gauss showed in the *Disquisitiones*. His proof may be clarified with the help of the concept of *content* of a polynomial in $\mathbb{Z}[x]$ – the gcd of its coefficients – as pointed out by Dedekind [1871], p.466. We denote the content of $f \in \mathbb{Z}[x]$ by $I(f)$.

Content Theorem. *If $g(x), h(x) \in \mathbb{Z}[x]$ then $I(g(x)h(x)) = I(g(x))I(h(x))$.*

Proof. By dividing each polynomial by its content we can reduce to the case where both $g(x)$ and $h(x)$ have content 1, in which case we want to prove that $I(g(x)h(x)) = 1$. Suppose

$$g(x) = b_rx^r + \cdots + b_1x + b_0, \quad h(x) = c_sx^s + \cdots + c_1x + c_0,$$

and suppose for the sake of contradiction that a prime number p divides $I(g(x)h(x))$, that is, p divides each coefficient of $g(x)h(x)$.

By hypothesis, there is a least i such that p does not divide b_i, and a least j such that p does not divide c_j. But then the coefficient of x^{i+j} in $g(x)h(x)$,

$$b_0c_{i+j} + b_1c_{i+j-1} + \cdots + b_ic_j + b_{i+1}c_{j-1} + \cdots + b_{i+j}c_0$$

is not divisible by p, since p divides all terms *except* b_ic_j. This contradiction shows that $I(g(x)h(x)) = 1$. □

This theorem has two pleasant corollaries, the first being Gauss's lemma itself, and the second a sharper unique factorisation theorem for $\mathbb{Z}[x]$ than we have for $\mathbb{Q}[x]$.

Gauss's Lemma. *If $g(x), h(x) \in \mathbb{Q}[x]$ are monic polynomials and $g(x)h(x) \in \mathbb{Z}[x]$ then $g(x), h(x) \in \mathbb{Z}[x]$.*

Proof. Multiply $g(x), h(x)$ by an integer m so that $mg(x), mh(x) \in \mathbb{Z}[x]$. Since $g(x)h(x) \in \mathbb{Z}[x]$ and $g(x)h(x)$ is monic, $I(m^2g(x)h(x)) = m^2$. On the other hand, $I(m^2g(x)h(x)) = I(mg(x))I(mh(x))$ by the content theorem. Thus

$$I(mg(x))I(mh(x)) = m^2.$$

But $I(mg(x))$ and $I(mh(x))$ divide the leading coefficient m of $mg(x)$ and $mh(x)$, hence both are $\leq m$. It follows that in fact $I(mg(x)) = I(mh(x)) = m$ and therefore $g(x), h(x) \in \mathbb{Z}[x]$. □

Unique Factorisation in $\mathbb{Z}[x]$. *The factorisation of a polynomial into irreducibles over $\mathbb{Z}$ is unique up to order and $\pm$ signs.*

Proof. Given $f(x) \in \mathbb{Z}[x]$ we first factorise $f(x)$ over $\mathbb{Q}[x]$ into

$$f(x) = rf_1(x)\cdots f_k(x) \text{ where } r \in \mathbb{Q} \text{ and } f_1, \ldots, f_k \text{ are monic.}$$

Then if we multiply each $f_i(x)$ by the lcm n_i of the denominators of its coefficients, and $r = m/n$ by its denominator n we get

$$nn_1\cdots n_kf(x) = mf_1^*(x)\cdots f_k^*(x), \qquad (*)$$

where each $f_i^*(x) = n_if_i(x)$ is a polynomial in $\mathbb{Z}[x]$ with content 1. Taking contents on both sides, it follows from the content theorem that

$$nn_1\cdots n_kI(f(x)) = m.$$

Since $I(f(x))$ is an integer, $nn_1\cdots n_k$ must divide m, with quotient $m^* \in \mathbb{Z}$, say. Thus we can divide both sides of the factorisation (*) in $\mathbb{Z}[x]$ to get

$$f(x) = m^*f_1^*(x)\cdots f_k^*(x)$$

– a factorisation of $f(x)$ in $\mathbb{Z}[x]$ differing from the irreducible factorisation in $\mathbb{Q}[x]$ only by constant factors. Since the latter is unique, up to constant factors, the same is true of factorisations in $\mathbb{Z}[x]$.

But if we now view a constant m^* as a zero degree polynomial of content m^*, it follows that factorisation in $\mathbb{Z}[x]$ is unique up to order and $\pm$ signs, being necessarily a product of an integer with polynomials of content 1. □

Exercise

4.6.1 Show that if $f \in \mathbb{Z}[x]$ has a factorisation $f = gh$ with $g, h \in \mathbb{Q}[x]$ then f factorises in $\mathbb{Z}[x]$ into rational multiples of g, h.

4.7 Eisenstein's Irreducibility Criterion

The question of irreducibility of integer polynomials over $\mathbb{Q}$ is not settled in practice by Gauss's lemma. A polynomial $f(x) \in \mathbb{Z}[x]$ still has infinitely many potential factors $g(x), h(x) \in \mathbb{Z}[x]$. One needs, say, a bound on the size of their coefficients in order to decide the existence of nontrivial factors in a finite number of steps. There are in fact ways of doing this, and they are a common feature of computer algebra systems. However, they are laborious to use by hand. An alternative is to use a criterion which does not apply in all cases, but which often gives the result at a glance. The best known of these is the following.

Eisenstein's Irreducibility Criterion. *If*

$$f(x) = x^n + a_{n-1}x^{n-1} + \cdots + a_1x + a_0 \in \mathbb{Z}[x]$$

and p is a prime that divides each a_i, and p^2 does not divide a_0, then $f(x)$ is irreducible over $\mathbb{Q}$.

Proof. Suppose on the contrary that there is a nontrivial factorisation $f(x) = g(x)h(x)$ where

$$g(x) = x^r + b_{r-1}x^{r-1} + \cdots + b_1x + b_0, \quad h(x) = x^s + c_{s-1}x^{s-1} + \cdots + c_1x + c_0.$$

By Gauss's lemma (Section 4.6) we can assume that $g(x), h(x) \in \mathbb{Z}[x]$. Now consider the polynomials $\bar{f}(x), \bar{g}(x), \bar{h}(x)$ that result from $f(x), g(x), h(x)$ by reducing their coefficients a_i, b_j, c_k mod p.

Since any relation between the a_i, b_j, c_k involving $+, -, \times$ yields the corresponding relation between their corresponding congruence classes $\bar{a}_i, \bar{b}_j, \bar{c}_k$ mod p, it follows from the definition of product of polynomials (Section 4.1) that $\bar{f}(x) = \bar{g}(x)\bar{h}(x)$.

This is a relation between polynomials with coefficients in $\mathbb{Z}/p\mathbb{Z}$, and $\mathbb{Z}/p\mathbb{Z}$ is a field (Section 2.7), hence unique factorisation of polynomials holds (Section 4.4). Since $\bar{f}(x) = x^n$, this means $\bar{g}(x) = x^r, \bar{h}(x) = x^s$. That is, all but the leading coefficients of $g(x)$ and $h(x)$ are congruent to 0 (mod p). In particular, $b_0 \equiv 0 \pmod{p}$, $c_0 \equiv 0 \pmod{p}$ and therefore $a_0 = b_0c_0$ is divisible by p^2, contrary to hypothesis. □

Eisenstein's irreducibility criterion shows immediately that $x^3 - 2$ and $x^5 - 2$ are irreducible over $\mathbb{Q}$ (take $p = 2$). We can also use it in certain cases where the nonleading coefficients have no prime divisor p, by suitable transformation of the polynomial. An important example is the pth cyclotomic polynomial, whose roots we saw in Section 3.7 to be vertices of the regular p-gon.

Theorem. *When p is prime the polynomial*

$$\Phi_p(z) = z^{p-1} + \cdots + z + 1$$

is irreducible over $\mathbb{Q}$.

Proof. Instead of $\Phi_p(z)$ we consider the polynomial $f(y) = \Phi_p(y+1)$, which is irreducible over $\mathbb{Q}$ if and only if $\Phi_p(z)$ is. Since

$$\Phi_p(z) = \frac{z^p - 1}{z - 1}$$

we have, by the binomial theorem,

$$\begin{aligned} f(y) &= \frac{(y+1)^p - 1}{y} \\ &= \frac{1}{y}\left[y^p + \binom{p}{1}y^{p-1} + \cdots + \binom{p}{p-1}y + 1 - 1\right] \\ &= y^{p-1} + \binom{p}{1}y^{p-2} + \cdots + \binom{p}{p-2}y + \binom{p}{p-1}, \end{aligned}$$

where $\binom{p}{i} = \frac{p(p-1)\cdots(p-i+1)}{i!}$ for $1 \le i \le p-1$ and in particular $\binom{p}{p-1} = p$.

Since $\binom{p}{i}$ is an integer it follows that its denominator $i!$ divides its numerator $p(p-1)\cdots(p-i+1)$. And since $i!$ does not have the factor p it follows by unique prime factorisation (Section 2.5) that the factor p of the numerator is a divisor of $\binom{p}{i}$. Since p^2 does *not* divide $\binom{p}{p-1} = p$, $f(y)$ satisfies Eisenstein's criterion and hence is irreducible over $\mathbb{Q}$, as required. □

Exercises

4.7.1 Transform $x^3 - 3x - 1$ into a polynomial which satisfies Eisenstein's criterion, and hence give another proof that $x^3 - 3x - 1 = 0$ has no rational solution (compare with Section 3.1).

4.7.2 Prove the following generalisation of Eisenstein's criterion. If $f(x) = a_n x^n + \cdots + a_1 x + a_0 \in \mathbb{Z}[x]$ and p is a prime that divides each a_i *except* a_n, and p^2 does not divide a_0, then $f(x)$ is irreducible over $\mathbb{Q}$.

4.7.3 Prove the "back-to-front" Eisenstein: if $f(x) = a_n x^n + \cdots + a_1 x + a_0 \in \mathbb{Z}[x]$ and p is a prime that divides each a_i except a_0, and p^2 does not divide a_n, then $f(x)$ is irreducible over $\mathbb{Q}$.

4.7.4 Show that the equation $y^3 + y^2 - 2y - 1 = 0$ for the regular heptagon (Exercise 1.3.3) is irreducible, by substituting $x + 2$ for y.

4.7.5 Show that $x^4 + 1$ is irreducible over $\mathbb{Q}$ by means of a suitable substitution, but that it factorises into a product of quadratics over $\mathbb{Q}(\sqrt{2})$.

4.7.6 Find the monic polynomial whose roots are $\pm\sqrt{2 \pm \sqrt{3}}$ and show that it is irreducible over $\mathbb{Q}$.

4.7.7 Find the minimal monic polynomial for $\frac{1}{5}\sqrt{50 - 10\sqrt{5}}$ (the side of the icosahedron in the unit sphere, see Section 1.9).

4.8* Cyclotomic Polynomials

The irreducibility proof for $\Phi_p(x) = x^{p-1} + \cdots + x + 1$ obviously works only when p is prime. As mentioned in Section 3.7, when n is not prime the polynomial $x^{n-1} + \cdots + x + 1$ factorises into nontrivial polynomials with integer coefficients, and the roots of one of these polynomials, $\Phi_n(x)$, are precisely the primitive n^{th} roots of unity. We shall now establish this property of $\Phi_n(x)$, the nth *cyclotomic polynomial*, leaving the proof of its irreducibility until Section 4.9*.

The factor theorem (Section 3.8 or 4.2) tells us that a polynomial is determined, up to a constant factor, by its roots (and their multiplicities, in the case of multiple roots). Hence there is exactly one monic polynomial with given roots.

Theorem. *The monic polynomial $\Phi_n(x)$ whose roots are the primitive n^{th} roots of unity has integer coefficients.*

Proof. Since the equation $x^n - 1 = 0$ has the n distinct roots $\zeta_n, \zeta_n^2, \ldots, \zeta_n^n = 1$, it follows by the factor theorem that

$$x^n - 1 = (x - \zeta_n)(x - \zeta_n^2)\cdots(x - \zeta_n^n). \qquad (*)$$

For each power ζ_n^i there is a least positive integer d such that $(\zeta_n^i)^d = 1$, namely the least d such that $id = kn$ for some k. Thus each ζ_n^i is a primitive d^{th} root of unity for some d. Also, $d|n$, because $\zeta_n^{id} = 1$ only if $id = kn$ for some k, and $\gcd(k, d) = 1$ because d is least. Conversely, for each divisor d of n, each dth root of unity is a power of ζ_n, in fact, a power of $\zeta_n^{i/d}$.

Thus the factors $(x - \zeta_n^i)$ of $x^n - 1$ in (*) can be partitioned into the products

$$\Phi_d(x) = \prod_{\zeta \text{ a primitive } d^{\text{th}} \text{ root of } 1} (x - \zeta)$$

for the divisors d of n, and hence

$$x^n - 1 = \prod_{d|n} \Phi_d(x).$$

It is clear from its definition that each $\Phi_d(x)$ is a monic polynomial.

We now argue by induction on d that each $\Phi_d(x)$ has integer coefficients. Since $\Phi_1(x) = x - 1$, obviously, this is true for $d = 1$. Supposing it true for each $d < n$, it remains to show that

$$\Phi_n(z) = (x^n - 1)/ \prod_{d|n, d<n} \Phi_d(x)$$

has integer coefficients.

We denote by $q(x)$ the polynomial $\prod_{d|n,d<n} \Phi_d(x)$, which is monic and with integer coefficients since the $\Phi_d(x)$ are (by induction). Letting

$$q(x) = x^r + q_{r-1}x^{r-1} + \cdots + q_1 x + q_0,$$
$$\Phi_n(x) = a_s x^s + a_{s-1}x^{s-1} + \cdots + a_1 x + a_0,$$

and using the equation

$$x^n - 1 = q(x)\Phi_n(x) = (x^r + \cdots + q_1 x + q_0)(a_s x^s + \cdots + a_1 x + a_0)$$

to solve successively for $a_0, a_1, \ldots, a_s$, we see $a_0, a_1, \ldots, a_s \in \mathbb{Q}$ and $a_s = 1$. But then it follows from Gauss's lemma that in fact $a_0, \ldots, a_s \in \mathbb{Z}$, which completes the induction. □

A few other properties of $\Phi_n(x)$ (and of the Euler ϕ function) fall out of this proof:

(i) The degree of $\Phi_n(x)$ is $\phi(n)$,

(ii) $\sum_{d|n} \phi(d) = n$, (compare with Section 2.9*),

(iii) The coefficients of $\Phi_n(x)$ can be calculated inductively, beginning with $\Phi_1(x) = x - 1$ and using the fact that

$$\Phi_n(x) = (x^n - 1) / \prod_{d|n, d<n} \Phi_d(x).$$

The first few cyclotomic polynomials are

$$\begin{aligned}
\Phi_2(x) &= x + 1, \\
\Phi_3(x) &= x^2 + x + 1, \\
\Phi_4(x) &= x^2 + 1, \\
\Phi_5(x) &= x^4 + x^3 + x^2 + x + 1, \\
\Phi_6(x) &= x^2 - x + 1, \\
\Phi_7(x) &= x^6 + x^5 + x^4 + x^3 + x^2 + x + 1, \\
\Phi_8(x) &= x^4 + 1, \\
\Phi_9(x) &= x^6 + x^3 + 1, \\
\Phi_{10}(x) &= x^4 - x^3 + x^2 - x + 1.
\end{aligned}$$

The coefficients are certainly integers; they even seem to be restricted to the values 0, 1 and -1. In fact this restriction is false, but it is a surprisingly long way to the first counterexample:

$$\Phi_{105}(x) = x^{48} + x^{47} + x^{46} - x^{43} - x^{42} - 2x^{41} - \cdots$$

Exercises

4.8.1 Prove statements (i) and (ii) above.

4.8.2 Check that the primitive 8th roots of unity are indeed the roots of $\Phi_8(x)$ as given above.

4.9* Irreducibility of Cyclotomic Polynomials

We do not know the coefficients of the cyclotomic polynomial $\Phi_n(x)$ well enough to prove its irreducibility via Eisenstein's criterion, except in the case where n is a prime p. However, we can use the idea from the *proof* of Eisenstein's criterion, reducing the coefficients of Φ_n modulo a prime p. As before, this drops us into the world $(\mathbb{Z}/p\mathbb{Z})[x]$ of polynomials with coefficients in the field $\mathbb{Z}/p\mathbb{Z}$, where unique factorisation rules. A proof of irreducibility can then be pushed through with the rather surprising intervention of calculus – or something like it – in the following lemma.

Lemma. *If p is a prime not dividing n, then $x^n - 1$ has no repeated factors in $(\mathbb{Z}/p\mathbb{Z})[x]$.*

Proof. For any polynomial

$$h(x) = a_r x^r + a_{r-1}x^{r-1} + \cdots + a_1 x + a_0$$

we define the *formal derivative* of $h(x)$ to be

$$Dh(x) = ra_r x^{r-1} + (r-1)a_{r-1}x^{r-2} + \cdots + a_1.$$

It can be checked directly from the definition (Exercise 4.9.1) that D satisfies the "product rule"

$$D(h_1(x)h_2(x)) = h_1(x) \cdot Dh_2(x) + Dh_1(x) \cdot h_2(x).$$

This implies that

$$D(f(x)^2 g(x)) = f(x)^2 Dg(x) + 2f(x)Df(x)g(x),$$

which has the divisor $f(x)$ in common with $f(x)^2 g(x)$. Thus *any $h(x)$ with a repeated factor $f(x)$ has that factor in common with $Dh(x)$.*

But $D(x^n - 1) = nx^{n-1} \neq 0$ in $(\mathbb{Z}/p\mathbb{Z})[x]$ since $p \nmid n$. Hence $x^n - 1$ and $D(x^n - 1)$ have no nontrivial common divisor (the only divisors of x^{n-1} are powers x^m, which do not divide $x^n - 1$), and therefore $x^n - 1$ has no repeated factor in $(\mathbb{Z}/p\mathbb{Z})[x]$. □

Our strategy now is to reduce $\Phi_n(x)$ modulo a prime p not dividing n, and show that a factorisation of $\Phi_n(x)$ implies a multiple factor of the reduced polynomial $\bar{\Phi}_n(x)$, and hence of $x^n - 1$. Another idea from Section 4.7 that assists in the proof is the fact that $(a+b)^p = a^p + b^p$ in $\mathbb{Z}/p\mathbb{Z}$, since p divides all the binomial coefficients $\binom{p}{i}$ for $1 \leq i \leq p-1$. It follows that if

$$\bar{g}(x) = a_r x^r + \cdots + a_1 x + a_0$$

then

$$\begin{aligned} \bar{g}(x)^p &= a_r^p x^{rp} + \cdots + a_1^p x^p + a_0^p \text{ in } (\mathbb{Z}/p\mathbb{Z})[x] \\ &= a_r x^{rp} + \cdots + a_1 x^p + a_0 \\ &= \bar{g}(x^p) \end{aligned}$$

since each $a_i^p = a_i$ in $\mathbb{Z}/p\mathbb{Z}$ by Fermat's little theorem (Section 2.8*).

Theorem. *For each $n \geq 1, \Phi_n(x)$ is irreducible over $\mathbb{Q}$.*

Proof. Suppose $\Phi_n(x) = f(x)g(x)$ where $f(x), g(x) \in \mathbb{Q}[x]$ are monic. Then in fact $f(x), g(x) \in \mathbb{Z}[x]$ by Gauss's lemma (Section 4.6). Without loss of generality we can assume $f(x)$ is nontrivial and satisfied by a primitive n^{th} root of unity, ζ. If p is a prime not dividing n then ζ^p is also a primitive n^{th} root of unity, and hence a root of $f(x)$ or $g(x)$.

If ζ^p is a root of $g(x)$ then ζ is a root of $g(x^p)$, hence $f(x)$ divides $g(x^p)$ in $\mathbb{Q}[x]$, by the general factor theorem of Section 4.2, hence in $\mathbb{Z}[x]$ by Gauss's lemma. Suppose

$$g(x^p) = f(x)h(x) \quad \text{where} \quad h(x) \in \mathbb{Z}[x].$$

Reducing this equation mod p (as in the proof of Eisenstein's criterion) we get

$$\bar{f}(x)\bar{h}(x) = \bar{g}(x^p) = \bar{g}(x)^p \quad \text{in} \quad (\mathbb{Z}/p\mathbb{Z})[x],$$

by the remark preceding the statement of the theorem. Since unique factorisation holds in $(\mathbb{Z}/p\mathbb{Z})[x]$ by Section 4.3, any irreducible factor of $\bar{f}(x)$ divides $\bar{g}(x)^p$ and hence $\bar{g}(x)$. It follows that $\bar{f}(x), \bar{g}(x)$ have a common factor in $(\mathbb{Z}/p\mathbb{Z})[x]$. If we now reduce the equation $\Phi_n(x) = f(x)g(x)$ mod p we get

$$\bar{\Phi}_n(x) = \bar{f}(x)\bar{g}(x),$$

which shows that $\bar{\Phi}_n(x)$ has a multiple factor in $(\mathbb{Z}/p\mathbb{Z})[x]$. But then, by Section 4.8, so has

$$x^n - 1 = \prod_{d|n} \bar{\Phi}_d(x),$$

contrary to the lemma.

Thus ζ^p is a root of $f(x)$ for each prime p not dividing n. Any m relatively prime to n is a product $m = p_1 p_2 \cdots p_k$ of such primes, and since the above argument shows $f(\xi) = 0 \Rightarrow f(\xi^{p_i}) = 0$ for *any* root ξ of $f(x)$, we get

$$f(\zeta) = 0 \Rightarrow f(\zeta^{p_1}) = 0 \Rightarrow f(\zeta^{p_1 p_2}) = 0 \Rightarrow \cdots \Rightarrow f(\zeta^m) = 0$$

for each of the $\phi(n)$ values of $m, 1 \leq m < n$, which are relatively prime to n. This means $f(x)$ has degree $\geq \phi(n)$ and hence $f(x) = \Phi_n(x)$, so $\Phi_n(x)$ is irreducible. □

Exercises

4.9.1 Check that $D(h_1(x)h_2(x)) = h_1(x) \cdot Dh_2(x) + Dh_1(x) \cdot h_2(x)$.

4.9.2 Use D to show

(i) $ax^2 + bx + c$ has a multiple root $\Leftrightarrow b^2 - 4ac = 0$,

(ii) $x^3 - px - q$ has a multiple root $\Leftrightarrow (p/3)^3 - (q/2)^2 = 0$ (compare with Exercise 1.6.4).

4.10 Discussion

The idea that a polynomial is just the sequence of its coefficients is not a modern one. In the late 16th century, when the first systems of algebraic symbolism were being tried, some authors avoided the use of a letter for the unknown by writing just the sequence of coefficients and, next to them, the corresponding exponents. For example, Bombelli [1572] would express the polynomial $2x^3+x+4$ by $2 \overset{3}{\smile} 1 \overset{1}{\smile} 4 \overset{0}{\smile}$, and Stevin [1585] would express it by 2③1①4⓪. Clearly, it remains only to list the coefficient 0 for any missing powers in order to dispense with exponents and represent the polynomial by its coefficients alone. Stevin also observed the analogy between polynomials and integers, and introduced the Euclidean algorithm for finding the gcd of two polynomials.

The creation of a root of $p(x)$ in the field $F[x]/p(x)F[x]$ is due to Kronecker [1887], though an interesting special case was given by Cauchy [1847]. Cauchy was unhappy about the definition $i = \sqrt{-1}$, and proposed instead to handle complex numbers as polynomials in $\mathbb{R}[x]/(x^2-1)\mathbb{R}[x]$. In other words, introduce an indeterminate x that satisfies $x^2 = -1$. All polynomials then reduce to the form $\alpha + \beta x$ with $\alpha, \beta \in \mathbb{R}$, and these expressions have the same algebraic behavior as $\alpha + i\beta$. In fact, if the polynomial $\alpha + \beta x$ is represented by the sequence (α, β) of its coefficients, one obtains a representation of complex numbers by ordered pairs of reals already given by Hamilton [1837].

Kronecker wanted to eliminate not only $\sqrt{-1}$ but irrational real numbers as well. He illustrated the process of elimination in his paper [1887] by the example

$$4(x^3 - a^3) \equiv (x-a)(2x+z+a)(2x-z+a) \pmod{z^2+3a^2}$$

"whereby the introduction of $\sqrt{-3}$ in the factorisation of $x^3 - a^3$ is avoided," as he put it. (Recall from Section 2.10 that Euler used this factorisation to prove that $x^3 + y^3 = z^3$ has no solution in positive integers.) Kronecker felt that this was the only way to study irreducible equations, because he did not believe in irrational numbers! He opposed the use of infinite sets, such as Dedekind cuts, and aimed to express all mathematical concepts in terms of finite sequences of integers. He is famous for saying: "God made the natural numbers, all the rest is the work of man." (According to the obituary by Weber [1893], Kronecker said this in a speech in 1886.) He is also said to have told Lindemann, à propos of the transcendence of π: "Of what use is your beautiful proof, since irrational numbers do not exist?"

As Kronecker probably realised, the reduction of mathematical concepts to finite sequences of integers can only be done in certain limited domains (which he happened to be interested in), such as algebraic number theory. Even then, it does not necessarily make the theory easier to understand. It means, for example, working with the congruence relation rather than with congruence classes. This is one reason Kronecker's work is harder to follow (for most people) than Dedekind's.

Actually, quite a lot was known about the arithmetic of irrational algebraic numbers even in ancient Greece. Enough to prove, for example, that the numbers

$a+b\sqrt{2}$, with $a, b \in \mathbb{Q}$, form a field. Book X of Euclid's *Elements* contains the key proposition that $1/(a+b\sqrt{c})$ is of the form $a'+b'\sqrt{c}$, along with many other propositions in the same "algebraic" vein. Of course these are lengths in Euclid, not numbers; the field concept is absent, and the proofs are entirely geometric. However, it is hard to believe that Euclid really thought this was geometry – more likely the results were just expressed that way because geometry was then the accepted language of mathematics. Sir Thomas Heath, the eminent Greek scholar and editor of the *Elements*, even speculated on the existence of a lost method by which such results were first discovered:

> That the Greeks must have had some analytical method which suggested the steps of such proofs seems certain, but what it was must remain apparently an insoluble mystery (Heath [1925], p.246).

Finding such a lost method is not out of the question. A lost method of Archimedes was found (amazingly) in 1906, revealing that Archimedes' geometric theorems on areas and volumes were actually discovered by reasoning about infinitesimals (Heath [1912]).

The first statement and proof of Gauss's lemma is in the *Disquisitiones*, article 42. Gauss uses it in article 341 to give the first proof that $x^{p-1}+\cdots+x+1$ is irreducible when p is prime. The simpler proof using Eisenstein's criterion is due to Eisenstein himself. He gave it as an illustration in the paper which introduces his criterion, Eisenstein [1850]. The main object of the paper is to prove irreducibility of the equation for *n-section of the lemniscate*, the analogue of the cyclotomic equation for the curve $(x^2+y^2)^2 = x^2-y^2$. This curve is called the lemniscate (from the Greek word for ribbon) because of its doubly-looped shape (Figure 4.10.1).

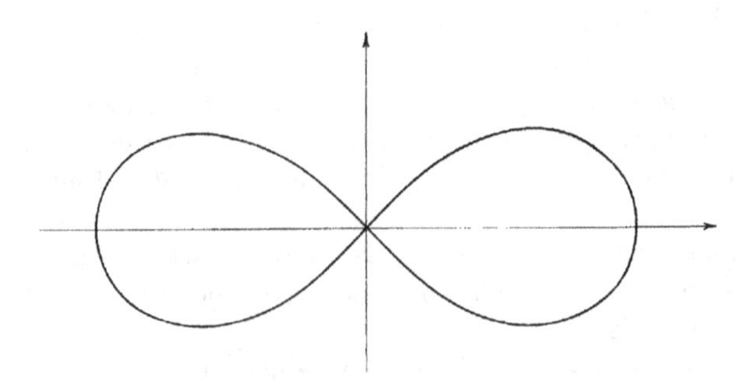

Fig. 4.10.1. The lemniscate

Jakob Bernoulli [1694] discovered that the lemniscate is analogous to the circle in the sense that its arc length is given by the integral of $1/\sqrt{1-x^4}$, whereas the arc length of the circle is given by the integral of $1/\sqrt{1-x^2}$. This analogy played an important role in the development of the theory of elliptic functions (see for example Stillwell [1989]), and Gauss hinted (*Disquisitiones*, article 335) at an analogy between n-section of the lemniscate and n-section of the circle. Abel [1827] took the hint and proved the astonishing theorem that n-section of the lemniscate is possible for *precisely the same* n as n-section of the circle.

The first proof of the irreducibility of $\Phi_n(x)$ was given by Kronecker [1854]. Dedekind [1857′] gave a much simpler proof, similar to the one given in Section 4.9*.

5 Fields

5.1 The Story So Far

We are now on the brink of solving several of the problems posed in Chapter 1, so it is a good time to review what we know. In Chapter 1 we looked at constructibility by straightedge and compass, and found that the constructible numbers result from 1 by closing under $+, -, \times, \div$ (hence they form a field) and square roots of positive numbers. At the same time, we found that the solution of certain geometric problems reduced to the solution of cubic or higher degree equations *not* obviously solvable by $+, -, \times, \div$ and square roots. In particular, duplication of the cube requires solution of the equation $x^3 - 2 = 0$, and construction of the regular p-gon requires solution of the equation $x^{p-1} + \cdots + x + 1 = 0$.

In Chapter 4 we found that the equations $h(x) = 0$ in question are *irreducible* over $\mathbb{Q}$, that is, $h(x)$ has no nontrivial factors in $\mathbb{Q}[x]$. This means, by Section 4.5, that $\mathbb{Q}[x]/h(x)\mathbb{Q}[x]$ is a field in which the indeterminate x behaves like a root α of the equation $h(x) = 0$. The elements of $\mathbb{Q}[x]/h(x)\mathbb{Q}[x]$ are represented by the polynomials $a_{n-1}x^{n-1} + \cdots + a_1 x + a_0$, where $a_0, a_1, \ldots, a_{n-1} \in \mathbb{Q}$ and $n = \text{degree}(h)$, since such polynomials are precisely the remainders that occur on division by $h(x)$. Since $h(\alpha) = 0$, the elements of $\mathbb{Q}(\alpha)$ are of the form $a_{n-1}\alpha^{n-1} + \cdots + a_1\alpha + a_0$ for the same reason.

For example, the field $\mathbb{Q}(\sqrt[3]{2})$ is the set of numbers $a_2(\sqrt[3]{2})^2 + a_1\sqrt[3]{2} + a_0$ where $a_0, a_1, a_2 \in \mathbb{Q}$, corresponding to the congruence classes in $\mathbb{Q}[x]/(x^3-2)\mathbb{Q}[x]$ represented by the polynomials $a_2x^2 + a_1x + a_0$. Thus $\mathbb{Q}(\sqrt[3]{2})$ is *three-dimensional over* $\mathbb{Q}$ in the sense that it takes three coordinates $a_0, a_1, a_2 \in \mathbb{Q}$ to specify a member of $\mathbb{Q}(\sqrt[3]{2})$. In the same sense, $\mathbb{Q}(\zeta_p)$ is $(p-1)$-dimensional over $\mathbb{Q}$ because ζ_p satisfies the irreducible equation $x^{p-1} + \cdots + x + 1 = 0$ of degree $p - 1$. Dimension will be defined precisely in Section 5.4.

The dimension of a field turns out to be the information we need to decide whether $\sqrt[3]{2}$ and ζ_p are constructible. We ask: can a field generated by square roots have dimension 3? (Or, in the case of ζ_p, can it have dimension $p-1$ for p prime?) The answer is easily obtained by a theorem of Dedekind (Section 5.5), which permits the calculation of dimension under quite general conditions.

5.2 Algebraic Numbers and Fields

The field $\mathbb{Q}(\alpha)$ obtained by adjoining a number α to $\mathbb{Q}$ was defined in Section 4.5 to be the closure of $\mathbb{Q} \cup \{\alpha\}$ under the operations $+, -, \times$ and $\div$ (by a nonzero number). Another description of $\mathbb{Q}(\alpha)$ is

$$\mathbb{Q}(\alpha) = \{f(\alpha)/g(\alpha) : f(\alpha), g(\alpha) \in \mathbb{Q}[\alpha] \text{ and } g(\alpha) \neq 0\},$$

that is, the quotients of polynomials in α with rational coefficients. On the one hand, polynomials $f(\alpha), g(\alpha)$ in α with rational coefficients certainly belong to the closure of $\mathbb{Q} \cup \{\alpha\}$, hence so does their quotient. Conversely, the set of quotients $f(\alpha)/g(\alpha)$, where $f(\alpha), g(\alpha) \in \mathbb{Q}[\alpha]$ and $g(\alpha) \neq 0$, is closed under $+, -, \times$ and $\div$ (by a nonzero number). These quotients are the *rational functions with coefficients in* $\mathbb{Q}$ (see also Section 3.4), *evaluated at* α.

It is known that certain numbers, such as π, do not satisfy any polynomial equation $h(x) = 0$ for $h(x) \in \mathbb{Q}[x]$. For such *transcendental* numbers α the description of $\mathbb{Q}(\alpha)$ as the rational functions of α is the simplest possible.

If, on the contrary, α satisfies an equation $h(x) = 0$ for some $h(x) \in \mathbb{Q}[x]$ then α is called an *algebraic* number, and $\mathbb{Q}(\alpha)$ can be described as a set of polynomials in α. Namely, suppose without loss of generality that $h(x)$ is irreducible over $\mathbb{Q}$ (we can always take an irreducible factor satisfied by α). Then it follows from the corollary in Section 4.5 that

$$\mathbb{Q}(\alpha) = \{a_{n-1}\alpha^{n-1} + \cdots + a_1\alpha + a_0 : a_0, a_1, \ldots, a_{n-1} \in \mathbb{Q}\} \qquad (*)$$

where $n = \text{degree}(h)$.

It is meaningful to define the *degree of* α *over* $\mathbb{Q}$ as $n = \text{degree}(h)$ because of the general factor theorem of Section 4.2, which shows that the irreducible polynomials satisfied by α differ only by constant factors. In particular, they all have the same degree, so the degree of α can mean only one thing.

For example, $\sqrt{2}$ has degree 2 over $\mathbb{Q}$ because $\sqrt{2}$ satisfies $x^2 - 2 = 0$ but (by the irrationality of $\sqrt{2}$) no linear equation with rational coefficients. Similarly, $\sqrt[3]{2}$ has degree 3 over $\mathbb{Q}$ by Exercise 4.5.1.

It turns out that degree is also a meaningful concept for the field $\mathbb{Q}(\alpha)$, though this is not so obvious since the number α defining $\mathbb{Q}(\alpha)$ is nowhere near unique. The meaningfulness of the degree of a field will be shown in Section 5.3.

Exercises

5.2.1 Show that $\mathbb{Q}(1 + \sqrt{2}) = \mathbb{Q}(\sqrt{2})$.

5.2.2 Show that $\sqrt{2}, \sqrt{3} \in \mathbb{Q}(\sqrt{2} + \sqrt{3})$.

5.2.3 Using the fact that $2\cos\frac{2\pi}{n} = \zeta_n + \zeta_n^{-1}$, or otherwise, show that all the numbers $\cos\frac{2\pi}{n}, \cos\frac{4\pi}{n}, \cos\frac{6\pi}{n}, \ldots$ belong to the field $\mathbb{Q}(\cos\frac{2\pi}{n})$.

5.3 Algebraic Elements over an Arbitrary Field

The idea of an algebraic number can be generalised to any field F by saying that α is *algebraic over* F if α satisfies an equation $g(x) = 0$ where $g \in F[x]$. If α is not algebraic over F then α is called *transcendental over* F. As in Section 5.2, any α which is algebraic over F has a well-defined *degree over* F, equal to the degree of any irreducible $g \in F[x]$ such that α satisfies $g(x) = 0$.

For example, $\sqrt[4]{2}$ has degree 2 over $\mathbb{Q}(\sqrt{2})$ because it satisfies $x^2 - \sqrt{2} = 0$ but does not satisfy any linear equation over $\mathbb{Q}(\sqrt{2})$ (Exercise 5.3.1).

The result $F(\alpha)$ of adjoining α to F is the closure of $F \cup \{\alpha\}$ under $+, -, \times, \div$, and we know from the corollary in Section 4.5 that

$$F(\alpha) = \{a_{n-1}\alpha^{n-1} + \cdots + a_1\alpha + a_0 : a_0, a_1, \ldots, a_{n-1} \in F\}$$

when n is the degree of α over F. We also know that α need not be a number and F need not be a number field. For any irreducible polynomial $g \in F[x]$ one can simultaneously create a root of $g(x) = 0$ and adjoin it to F by constructing the field $F[x]/g(x)F[x]$, in which x itself satisfies $g(x) = 0$ by definition (Section 4.5).

A further useful generalisation is $F(\alpha_1, \ldots, \alpha_n)$, which can be defined as $F(\alpha_1)\ldots(\alpha_n)$, that is, the result of adjoining $\alpha_1, \ldots, \alpha_n$ to F in that order. The result is actually independent of the order, by the following theorem.

Theorem. *$F(\alpha_1, \ldots, \alpha_n)$ is the closure of $F \cup \{\alpha_1, \ldots, \alpha_n\}$ under $+, -, \times, \div$.*

Proof. We argue by induction on n. The result is true by definition for $n = 1$, so assume inductively that

$$F(\alpha_1, \ldots, \alpha_k) = \text{closure}(F \cup \{\alpha_1, \ldots, \alpha_k\}).$$

Then

$$\begin{aligned} F(\alpha_1, \ldots, \alpha_{k+1}) &= F(\alpha_1, \ldots, \alpha_k)(\alpha_{k+1}) \quad \text{by definition} \\ &= \text{closure}(F(\alpha_1, \ldots, \alpha_k) \cup \{\alpha_{k+1}\}) \quad \text{by } n = 1 \text{ case} \\ &= \text{closure}(\text{closure}(F \cup \{\alpha_1, \ldots, \alpha_k\}) \cup \{\alpha_{k+1}\}) \quad \text{by induction} \\ &= \text{closure}(F \cup \{\alpha_1, \ldots, \alpha_{k+1}\}), \end{aligned}$$

which completes the induction. □

Exercises

5.3.1 Show that $\sqrt[4]{2} \notin \mathbb{Q}(\sqrt{2})$.

5.3.2 Show that $\mathbb{Q}(\sqrt{2} + \sqrt{3}) = \mathbb{Q}(\sqrt{2}, \sqrt{3})$ (compare with Exercise 5.2.2).

5.4 Degree of a Field over a Subfield

The degree of a field E over a field F is defined to be the dimension of E as a vector space over F. Thus some readers will be able to relax and lean on their knowledge of linear algebra at this point. However, for those who do not have such knowledge, the following is a self-contained treatment of the tiny amount of linear algebra we actually need. (Even experts may be interested to note that we are applying linear algebra to the highly *non*linear algebra of algebraic numbers.)

If E, F are fields with $E \supseteq F$ we call E an *extension* (field) of F and call F a *subfield* of E. E is said to be *finite-dimensional over* F if there are elements $\epsilon_1, \ldots, \epsilon_n \in E$ such that any $\epsilon \in E$ has the form

$$\epsilon = f_1\epsilon_1 + \cdots + f_n\epsilon_n \quad \text{where} \quad f_1, \ldots, f_n \in F.$$

The elements $\epsilon_1, \ldots, \epsilon_n$ are said to *span* E *over* F.

For example, the elements 1, $\sqrt{2}$ span $\mathbb{Q}(\sqrt{2})$ over $\mathbb{Q}$ because each member of $\mathbb{Q}(\sqrt{2})$ has the form $a + b\sqrt{2}$ with $a, b \in \mathbb{Q}$, by Section 4.5. More generally, *if α is an algebraic number of degree n then $1, \alpha, \ldots, \alpha^{n-1}$ span $\mathbb{Q}(\alpha)$ over $\mathbb{Q}$* because each member of $\mathbb{Q}(\alpha)$ has the form $a_0 + a_1\alpha + \cdots + a_{n-1}\alpha^{n-1}$ where $a_0, a_1, \ldots, a_{n-1} \in \mathbb{Q}$, also by Section 4.5.

Elements $\epsilon_1, \ldots, \epsilon_n \in E$ are said to be *independent* over F if there are no $f_1, \ldots, f_n \in F$, not all zero, such that

$$f_1\epsilon_1 + \cdots + f_n\epsilon_n = 0.$$

Otherwise $\epsilon_1, \ldots, \epsilon_n$ are said to be *dependent* over F.

For example, $1, \alpha, \ldots, \alpha^{n-1} \in \mathbb{Q}(\alpha)$ are independent over $\mathbb{Q}$ because if $a_0, a_1, \ldots, a_{n-1} \in \mathbb{Q}$ are not all zero then $a_0 + a_1\alpha + \cdots + a_{n-1}\alpha^{n-1} \neq 0$ by definition of the degree n of α (Section 5.1). In the special case of $\mathbb{Q}(\sqrt{2})$ the independence of $1, \sqrt{2}$ is equivalent to the irrationality of $\sqrt{2}$ (see Exercise 5.4.1).

An independent spanning set $\epsilon_1, \ldots, \epsilon_n$ is called a *basis* for E *over* F. Thus *if α is of degree n then $1, \alpha, \ldots, \alpha^{n-1}$ is a basis for $\mathbb{Q}(\alpha)$ over $\mathbb{Q}$*. The key property of bases is that they all have the same size, as we shall prove below using the following lemma. It is here that the field properties of F are crucial, inasmuch as they permit the solution of sets of linear equations.

Lemma. *A system of equations*

$$a_{i1}x_1 + \cdots + a_{in}x_n = 0, \quad i = 1, \ldots, m,$$

where $m < n$ and the $a_{ij} \in F$, has a nonzero solution $x_1, \ldots, x_n \in F$.

Proof. We argue by induction on m, the case $m = 0$ being trivial. Suppose then that the lemma is true for $m < k$ and that we are given

$$a_{i1}x_1 + \cdots + a_{in}x_n = 0 \text{ for } i = 1, \ldots, k \qquad (k)$$

with $k < n$. The equations (k) certainly have a nonzero solution $x_1, \ldots, x_n \in F$ if all $a_{ij} = 0$.

If some $a_{ij} \neq 0$ we can assume $a_{11} \neq 0$ by suitable renaming of unknowns and reordering of equations. We can then reduce the system (k) to the form

$$\begin{aligned} a_{11}x_1 + a_{12}x_2 + \cdots + a_{1n}x_n &= 0 \\ b_{22}x_2 + \cdots + b_{2n}x_n &= 0 \\ &\vdots \\ b_{k2}x_2 + \cdots + b_{kn}x_n &= 0 \end{aligned} \qquad (k-1)$$

by replacing the $i^{\text{th}} (i > 1)$ equation of (k) by (1^{st}equation $- \frac{a_{11}}{a_{i1}} \times i^{\text{th}}$equation) if $a_{i1} \neq 0$, and otherwise leaving it unchanged.

The last $(k-1)$ equations of system $(k-1)$ have a nonzero solution $x_2, \ldots, x_n \in F$ by induction, and we can substitute this solution in the first equation and solve for $x_1 \in F$ since $a_{11} \neq 0$. This completes the induction. □

Theorem. *Any two bases of E over F have the same number of elements.*

Proof. It will suffice to show that if $\epsilon_1, \ldots, \epsilon_n$ is a basis for E over F then any $\alpha_1, \ldots, \alpha_{n+1} \in E$ are dependent, because this implies that one basis cannot have more elements than another.

Given $\alpha_1, \ldots, \alpha_{n+1} \in E$, suppose that they have the following representations in terms of the basis elements.

$$\alpha_i = a_{i1}\epsilon_1 + \cdots + a_{in}\epsilon_n \quad \text{where} \quad a_{i1}, \ldots, a_{in} \in F. \tag{1}$$

We wish to find $x_1, \ldots, x_{n+1} \in F$, not all zero, such that

$$\alpha_1 x_1 + \cdots + \alpha_{n+1} x_{n+1} = 0. \tag{2}$$

By (1), this is the same as

$$(a_{11}\epsilon_1 + \cdots + a_{1n}\epsilon_n)x_1 + \cdots + (a_{n+1,1}\epsilon_1 + \cdots + a_{n+1,n}\epsilon_n)x_{n+1} = 0, \tag{3}$$

or

$$(a_{11}x_1 + \cdots + a_{n+1,1}x_{n+1})\epsilon_1 + \cdots + (a_{1n}x_1 + \cdots + a_{n+1,n}x_{n+1})\epsilon_n = 0. \tag{4}$$

Since $\epsilon_1, \ldots, \epsilon_n$ are independent, their respective coefficients in (4) must all be zero. Thus $x_1, \ldots, x_{n+1}$ have to satisfy the n equations

$$a_{1i}x_1 + \cdots + a_{n+1,i}x_{n+1} = 0 \quad \text{for} \quad i = 1, \ldots, n. \tag{5}$$

Since there are more unknowns than equations, the lemma gives us a solution $x_1, \ldots, x_{n+1} \in F$, not all zero, as required for the dependence of $\alpha_1, \ldots, \alpha_{n+1}$. □

Thus it is meaningful to define the *dimension* of E over F as the size of any basis for E over F. It is denoted by $(E : F)$. In particular, *if α is an algebraic number of degree n then* $(\mathbb{Q}(\alpha) : \mathbb{Q}) = n$ because $\mathbb{Q}(\alpha)$ has the basis $1, \alpha, \ldots, \alpha^{n-1}$ over $\mathbb{Q}$. Motivated by this example, the dimension of *any* field E over F is also called the *degree* of E over F. Finally, we also say that an extension of finite degree is *algebraic* because of the following corollary to the theorem.

Corollary. *If $(E : F) = n$ then each $\epsilon \in E$ is algebraic (of degree $\leq n$) over F.*

Proof. Since any $n + 1$ elements of E are dependent, this is true in particular of $1, \epsilon, \ldots, \epsilon^n$. Hence they satisfy an equation

$$a_0 + a_1\epsilon + \cdots + a_n\epsilon^n = 0$$

where $a_0, \ldots, a_n \in F$ are not all zero. In other words, ϵ satisfies $g(x) = 0$ where $g(x) \in F[x]$ is of degree $\leq n$. □

This concludes our treatment of linear algebra. Readers who already knew it should now resume paying attention. We are about to study the effect of *varying the base field F* – something not usually done in linear algebra, but crucial in the theory of fields.

Exercises

5.4.1 Show that 1, $\sqrt{2}$ are independent over $\mathbb{Q}$.

5.4.2 Show that 1, $\sqrt{3}$ are independent over $\mathbb{Q}(\sqrt{2})$.

5.5 Degree of an Iterated Extension

The result $E = F(\alpha_1, \dots, \alpha_k)$ of adjoining several elements $\alpha_1, \dots, \alpha_k$ to a field F appears very difficult to compute. For example, what is the degree of $\mathbb{Q}(\sqrt[3]{2}, \sqrt[5]{3})$? It is not even clear that the degree is finite, since the element $\sqrt[3]{2} + \sqrt[5]{3}$ in particular is not obviously the root of a polynomial with rational coefficients. Such questions may be answered by viewing $F(\alpha_1, \dots, \alpha_k)$ as the *iterated extension* $F(\alpha_1)\dots(\alpha_k)$ (as it was originally defined, in Section 5.3), and considering the extensions by α_1, by $\alpha_2, \dots$, by α_k separately. Then the problem is reduced to computing the combined effect of two field extensions, say from F to B (B for "between") and from B to E.

The effect of two extensions is captured by the following theorem.

Dedekind Product Theorem. *If $E \supseteq B \supseteq F$ are fields with $\epsilon_1, \dots, \epsilon_m \in E$ a basis for E over B and $\beta_1, \dots, \beta_n \in B$ a basis for B over F, then the mn products $\epsilon_i\beta_j$ form a basis for E over F. In particular,*

$$(E : F) = mn = (E : B)(B : F).$$

Proof. Since any $\epsilon \in E$ is expressible in the form

$$\epsilon = b_1\epsilon_1 + \cdots + b_m\epsilon_m \quad \text{where} \quad b_1, \dots, b_m \in B,$$

and each $b_i \in B$ is expressible in the form

$$b_i = f_{i1}\beta_1 + \cdots + f_{in}\beta_n \quad \text{where} \quad f_{i1}, \dots, f_{in} \in F,$$

we have

$$\epsilon = (f_{11}\beta_1 + \cdots + f_{1n}\beta_n)\epsilon_1 + \cdots + (f_{m1}\beta_1 + \cdots + f_{mn}\beta_n)\epsilon_m,$$

and hence the products $\epsilon_i\beta_j$ span E over F.

To show that they are independent over F, suppose there are $a_{ij} \in F$ with

$$\begin{aligned} 0 &= \sum_{i,j} a_{ij}\epsilon_i\beta_j \\ &= (a_{11}\beta_1 + \cdots + a_{1n}\beta_n)\epsilon_1 + \cdots + (a_{m1}\beta_1 + \cdots + a_{mn}\beta_n)\epsilon_m. \end{aligned}$$

Since ϵ_i has coefficient $a_{i1}\beta_1 + \cdots + a_{in}\beta_n \in B$, it follows from the independence of $\epsilon_1, \dots, \epsilon_m$ over B that

$$a_{i1}\beta_1 + \cdots + a_{in}\beta_n = 0 \quad \text{for each} \quad i.$$

But then it follows from the independence of $\beta_1, \dots, \beta_n$ over F that each $a_{ij} = 0$, as required. □

This theorem gives a surprisingly easy way to answer questions about the degree of sums of surds, such as those posed at the beginning of this section. It is immediate that the degree is finite, and its exact value can be obtained with the help of irreducibility information.

Example. $\mathbb{Q}(\sqrt[3]{2}, \sqrt[5]{3})$.

Viewing $\mathbb{Q}(\sqrt[3]{2}, \sqrt[5]{3})$ as the result of two extensions, $\mathbb{Q}$ to $\mathbb{Q}(\sqrt[5]{3})$ and $\mathbb{Q}(\sqrt[5]{3})$ to $\mathbb{Q}(\sqrt[5]{3})(\sqrt[3]{2})$, we observe:

(i) $(\mathbb{Q}(\sqrt[5]{3}) : \mathbb{Q}) = 5$ because $\sqrt[5]{3}$ satisfies $x^5 - 3 = 0$ and $x^5 - 3$ is irreducible over $\mathbb{Q}$ (for example by the Eisenstein irreducibility criterion, Section 4.7).

(ii) Any $\alpha \in \mathbb{Q}(\sqrt[5]{3})$ has degree 1 or 5 over $\mathbb{Q}$ since

$$\begin{aligned}\mathbb{Q}(\sqrt[5]{3}) \supseteq \mathbb{Q}(\alpha) \supseteq \mathbb{Q} &\Rightarrow (\mathbb{Q}(\sqrt[5]{3}) : \mathbb{Q}) = (\mathbb{Q}(\sqrt[5]{3}) : \mathbb{Q}(\alpha))(\mathbb{Q}(\alpha) : \mathbb{Q}) \\ &\Rightarrow (\mathbb{Q}(\alpha) : \mathbb{Q}) = 1 \text{ or } 5 \text{ (the only divisors of 5).}\end{aligned}$$

(iii) $x^3 - 2$ is irreducible over $\mathbb{Q}$, hence none of its roots are in $\mathbb{Q}(\sqrt[5]{3})$ by (ii). This implies $x^3 - 2$ is also irreducible over $\mathbb{Q}(\sqrt[5]{3})$, since a factorisation would have a linear factor, implying a root in $\mathbb{Q}(\sqrt[5]{3})$. Thus $(\mathbb{Q}(\sqrt[5]{3})(\sqrt[3]{2}) : \mathbb{Q}(\sqrt[5]{3})) = 3$.

(iv) $(\mathbb{Q}(\sqrt[3]{2}, \sqrt[5]{3}) : \mathbb{Q}) = 5 \times 3 = 15$ by (i) and (iii).

(v) Hence any element of $\mathbb{Q}(\sqrt[3]{2}, \sqrt[5]{3})$, in particular $\sqrt[3]{2} + \sqrt[5]{3}$, has degree ≤ 15 by Corollary 5.4. □

This last result is a special case of the following corollary to the theorem.

Corollary. *If α, β are algebraic numbers then so are $\alpha + \beta, \alpha - \beta, \alpha\beta$ and α/β (for $\beta \neq 0$).*

Proof. Suppose degree$(\alpha) = m$ and degree$(\beta) = n$. All the numbers in question belong to $\mathbb{Q}(\alpha, \beta)$ and

$$\begin{aligned}(\mathbb{Q}(\alpha, \beta) : \mathbb{Q}) &= (\mathbb{Q}(\alpha, \beta) : \mathbb{Q}(\beta))(\mathbb{Q}(\beta) : \mathbb{Q}) \quad \text{by the theorem} \\ &\leq mn\end{aligned}$$

since $(\mathbb{Q}(\alpha, \beta) : \mathbb{Q}(\beta)) \leq \text{degree}(\alpha) = m$ and $(\mathbb{Q}(\beta) : \mathbb{Q}) = \text{degree}(\beta) = n$. Thus all members of $\mathbb{Q}(\alpha, \beta)$ are of degree $\leq mn$ by Corollary 5.4, and hence are algebraic. □

Remark. By applying the Dedekind product theorem to iterated extensions, and varying the order in which elements are adjoined, we can sometimes establish irreducibility of polynomials over fields other than $\mathbb{Q}$. The above example does this, in a roundabout way, in showing that $x^3 - 2$ is irreducible over $\mathbb{Q}(\sqrt[5]{3})$. A more telling example is the field $\mathbb{Q}(\sqrt[4]{2}, i)$. On the one hand we have

$$(\mathbb{Q}(\sqrt[4]{2}, i) : \mathbb{Q}) = (\mathbb{Q}(\sqrt[4]{2}, i) : \mathbb{Q}(\sqrt[4]{2}))(\mathbb{Q}(\sqrt[4]{2}) : \mathbb{Q})$$

with

$$(\mathbb{Q}(\sqrt[4]{2}, i) : \mathbb{Q}(\sqrt[4]{2})) = 2,$$

because i does not belong to the real field $\mathbb{Q}(\sqrt[4]{2})$, and

$$(\mathbb{Q}(\sqrt[4]{2}) : \mathbb{Q}) = 4,$$

because $x^4 - 2$ is irreducible over $\mathbb{Q}$ by Eisenstein's criterion. On the other hand we have

$$\begin{aligned}(\mathbb{Q}(\sqrt[4]{2}, i) : \mathbb{Q}) &= (\mathbb{Q}(\sqrt[4]{2}, i) : \mathbb{Q}(i))(\mathbb{Q}(i) : \mathbb{Q}) \\ &= (\mathbb{Q}(\sqrt[4]{2}, i) : \mathbb{Q}(i)) \times 2\end{aligned}$$

and hence

$$(\mathbb{Q}(\sqrt[4]{2}, i) : \mathbb{Q}(i)) = 4.$$

Thus $\sqrt[4]{2}$ has degree 4 over $\mathbb{Q}(i)$ and hence $x^4 - 2$ is also irreducible over $\mathbb{Q}(i)$.

Exercises

5.5.1 Deduce from Exercise 5.4.2 that $(\mathbb{Q}(\sqrt{2}, \sqrt{3}) : \mathbb{Q}(\sqrt{2})) = 2$ and hence that $(\mathbb{Q}(\sqrt{2}, \sqrt{3}) : \mathbb{Q}) = 4$.

5.5.2 Deduce from Exercises 5.5.1 and 5.4.2 that $\sqrt{2} + \sqrt{3}$ has degree ≤ 4, and find a fourth degree polynomial in $\mathbb{Q}[x]$ satisfied by it.

5.5.3 Deduce from Exercises 5.5.1 and 5.3.2 that the polynomial found in exercise 5.5.2 is irreducible. Also see whether you can prove this directly, using Eisenstein's criterion.

5.5.4 Show that $(\mathbb{Q}(\zeta_n) : \mathbb{Q}(\cos \frac{2\pi}{n})) = 2$ for $n > 2$, and hence show that the degree of $\cos \frac{2\pi}{n}$ is $\phi(n)/2$. Use this to give new proofs (different from Exercises 4.7.1 and 4.7.4) that $\cos \frac{2\pi}{7}$ and $\cos \frac{\pi}{9}$ have degree 3.

5.5.5 For which integers n is $\cos \frac{2\pi}{n}$ rational?

5.6 Degree of Constructible Numbers

We are now poised to show that the classical construction problems are unsolvable. We know from Section 1.3 that any constructible number α is obtained from rational numbers by $+, -, \times, \div$ and $\sqrt{}$, hence α lies in a field $\mathbb{Q}(\alpha_1, \ldots, \alpha_n)$ where α_1 is the square root of a rational number and each α_{i+1} is the square root of an element of $\mathbb{Q}(\alpha_1, \ldots, \alpha_i)$. This puts a severe restriction on the degree of α, as the following theorem shows.

Theorem. *If $\alpha_1, \ldots, \alpha_n$ are numbers with $\alpha_1^2 \in \mathbb{Q}$ and $\alpha_{i+1}^2 \in \mathbb{Q}(\alpha_1, \ldots, \alpha_i)$ for each i, then the degree of $\mathbb{Q}(\alpha_1, \ldots, \alpha_n)$ is a power of 2. The degree of any $\alpha \in \mathbb{Q}(\alpha_1, \ldots, \alpha_n)$ is also a power of 2.*

Proof. Since each α_i satisfies a quadratic equation over $\mathbb{Q}(\alpha_1, \ldots, \alpha_{i-1})$ (over $\mathbb{Q}$ when $i = 1$) we have

$$(\mathbb{Q}(\alpha_1, \ldots, \alpha_i) : \mathbb{Q}(\alpha_1, \ldots, \alpha_{i-1})) = 1 \text{ or } 2 \quad \text{for each } i.$$

Hence by the Dedekind product theorem (Section 5.5)

$$\begin{aligned}(\mathbb{Q}(\alpha_1,\ldots,\alpha_n):\mathbb{Q}) &= (\mathbb{Q}(\alpha_1,\ldots,\alpha_n):\mathbb{Q}(\alpha_1,\ldots,\alpha_{n-1}))\\ &\quad\times(\mathbb{Q}(\alpha_1,\ldots,\alpha_{n-1}):\mathbb{Q}(\alpha_1,\ldots,\alpha_{n-2}))\\ &\quad\vdots\\ &\quad\times(\mathbb{Q}(\alpha_1):\mathbb{Q})\\ &= 2^m \quad \text{for some natural number } m\le n.\end{aligned}$$

If $\alpha\in\mathbb{Q}(\alpha_1,\ldots,\alpha_n)$ the Dedekind product theorem also gives

$$2^m=(\mathbb{Q}(\alpha_1,\ldots,\alpha_n):\mathbb{Q})=(\mathbb{Q}(\alpha_1,\ldots,\alpha_n):\mathbb{Q}(\alpha))(\mathbb{Q}(\alpha):\mathbb{Q}).$$

Thus $(\mathbb{Q}(\alpha):\mathbb{Q})$ is a divisor of 2^m and hence also a power of 2. □

Corollary 1. *The problem of duplication of the cube is unsolvable.*

Proof. As we know from Section 1.5, this problem is equivalent to construction of the number $\sqrt[3]{2}$. However, $\sqrt[3]{2}$ has degree 3 by the irreducibility of x^3-2 over $\mathbb{Q}$ (Section 4.7), and 3 is not a power of 2, so $\sqrt[3]{2}$ is not constructible. □

Corollary 2. *The problem of trisection of the angle is unsolvable.*

Proof. We know from Section 1.5 that trisection of the special (constructible) angle $\pi/3$ is equivalent to construction of the root $\cos\frac{\pi}{9}$ of $8x^3-6x-1=0$. The reducibility of this equation over $\mathbb{Q}$ is equivalent to that of the equation $y^3-3y-1=0$ satisfied by $y=2x$, which is in turn equivalent to that of the equation $z^3+3z^2-3=0$ satisfied by $z=y-1$, and the latter is evidently irreducible by Eisenstein's criterion (Section 4.7). Thus $\cos\frac{\pi}{9}$ has degree 3, and it is therefore not constructible. □

Corollary 3. *If p is prime then the regular p-gon is constructible only if $p-1$ is a power of 2.*

Proof. We know from Sections 1.5 and 3.7 that construction of the regular p-gon is equivalent to construction of a solution of the equation

$$z^{p-1}+\cdots+z+1=0.$$

We also know from Section 4.7 that this equation is irreducible when p is prime. Thus its solutions are of degree $p-1$, and hence constructible only if $p-1$ is a power of 2. □

It was shown by Gauss [1801], article 365, that the regular p-gon is indeed constructible whenever $p-1$ is a power of 2. This result is easier to prove when the concepts of group theory have been brought into play, and we shall therefore postpone it until Chapter 9. Primes p for which $p-1$ is a power of 2 are actually of the form $2^{2^h}+1$ (see Exercises 5.6.1 and 5.6.2) and are called *Fermat* primes. Only five of them are known, namely 3, 5, 17, 257, 65537, corresponding to $h=0,1,2,3,4$.

The connection between primes $p = 2^{2^h} + 1$ and regular p-gons was not known to Fermat, and in fact his interest in these primes was based on a mistake. Observing that $2^{2^h} + 1$ is prime for $h = 0, 1, 2, 3, 4$, Fermat [1640] conjectured that $2^{2^h} + 1$ is prime for *all* natural numbers h, thus realising the dream of a simple function whose values are all prime. Alas, Euler [1738] found that 641 divides $2^{2^5} + 1$, and since then divisors of many more numbers $2^{2^h} + 1$ have been found. It may well be that there no more prime values of $2^{2^h} + 1$ whatever. Yet Fermat was onto an interesting source of primes all the same. It is quite easy to show (Exercise 5.6.3) that $\gcd(2^{2^m} + 1, 2^{2^n} + 1) = 1$ when $m \neq n$, hence *the prime divisors of* $2^{2^m} + 1$ *are different from those of any other* $2^{2^n} + 1$. Amongst other things, this gives a new proof that there infinitely many primes (Goldbach [1730]).

Exercises

5.6.1 Show that $x^{2n+1} + 1 = (x + 1)(x^{2n} - x^{2n-1} + \cdots - x + 1)$.

5.6.2 Deduce from Exercise 5.6.1 that $2^k + 1$ is not prime if k has an odd divisor $2n + 1$, hence that $2^k + 1$ is prime only if $k = 2^h$.

5.6.3 Assuming that a prime p divides $2^{2^m} + 1$, so $2^{2^m} \equiv -1 \pmod{p}$, deduce that $2^{2^n} \equiv 1 \pmod{p}$ for $n > m$, and hence that p does not divide $2^{2^n} + 1$.

5.7* Regular n-gons

Theorem 5.6 also enables us to say that a regular n-gon is constructible only if $\phi(n)$ is a power of 2, since a primitive n^{th} root of unity has degree $\phi(n)$ by the irreducibility of the cyclotomic polynomial $\Phi_n(x)$ (Section 4.9*). However, this is not very informative until we know how to compute $\phi(n)$.

Lemma. *If* $n = p_1^{i_1} p_2^{i_2} \cdots p_k^{i_k}$ *is the prime factorisation of* n *then*

$$\phi(n) = p_1^{i_1 - 1}(p_1 - 1) \cdots p_k^{i_k - 1}(p_k - 1).$$

Proof. Since the factors $p_1^{i_1}, p_2^{i_2}, \ldots, p_k^{i_k}$ are relatively prime, it follows from Section 2.9* that

$$\phi(n) = \phi(p_1^{i_1}) \cdots \phi(p_k^{i_k}),$$

and hence it remains to evaluate $\phi(p^i)$ for a prime p. There are p^i positive integers $m \leq p^i$, and of these only the p^{i-1} multiples of p have a nontrivial divisor in common with p^i. Hence

$$\phi(p^i) = p^i - p^{i-1} = p^{i-1}(p - 1),$$

and the lemma follows. □

We can now decide when $\phi(n)$ is a power of 2, and hence when the regular n-gon is constructible.

Theorem. *The regular n-gon is constructible only if n is of the form*

$$n = 2^m p_1 p_2 \cdots p_k$$

where $p_1, p_2, \ldots, p_k$ are distinct Fermat primes.

Proof. Suppose $n = p_0^{i_0} p_1^{i_1} \cdots p_k^{i_k}$ is the prime factorisation of n with $p_0 < p_1 < \cdots < p_k$. Then

$$\phi(n) = p_0^{i_0-1}(p_0 - 1)p_1^{i_1-1}(p_1 - 1) \cdots p_k^{i_k-1}(p_k - 1)$$

by the lemma, and if the regular n-gon is constructible there is an $l \in \mathbb{N}$ such that

$$2^l = p_0^{i_0-1}(p_0 - 1)p_1^{i_1-1}(p_1 - 1) \cdots p_k^{i_k-1}(p_k - 1).$$

It follows by unique prime factorisation in $\mathbb{N}$ (Section 2.5) that

(i) p_0 has a nonzero exponent $i_0 - 1 = m$ only if $p_0 = 2$,

(ii) the remaining p_j have exponents $i_j - 1 = 0$,

(iii) each of $p_1 - 1, \ldots, p_k - 1$ is a power of 2, that is, $p_1, \ldots, p_k$ are distinct Fermat primes. □

It is easy to see (Exercise 5.7.1) that if the regular p_i-gons are constructible then so is the regular $2^m p_1 \cdots p_k$-gon. Thus the converse of the theorem – that the regular n-gon *is* constructible if $n = 2^m p_1 \cdots p_k$ for distinct Fermat primes $p_1, \ldots, p_k$ – reduces to the constructibility of the regular p-gon for each Fermat prime p. As mentioned in Section 5.6, we shall establish the latter construction in Chapter 9. From it we shall deduce that the n-section of an arbitrary angle is constructible if and only if $n = 2^m$, generalising the argument against trisection used in Section 5.6.

Exercises

5.7.1 Show that if $\gcd(p, q) = 1$ and the regular p-gon and q-gon are constructible, then so is the regular pq-gon, and also the regular $2^m pq$-gon.

5.7.2 Use Exercise 5.5.4 and the Lemma 5.7* to show that $\cos \frac{2\pi}{n}$ is of degree 3 only for $n = 7, 14, 9, 18$. Also find the n for which $\cos \frac{2\pi}{n}$ is of degree 4.

5.8 Discussion

Wantzel [1837] proved that duplication of the cube and trisection of the angle by straightedge and compass are impossible. These historic results can actually be proved by quite elementary methods. For example, it can be proved that $\sqrt[3]{2}$ is not expressible in terms of rational numbers and square roots by induction on the depth of nesting of $\sqrt{\ }$ signs. According to Bieberbach [1952], p.123, the number theorist Edmund Landau (1877-1938) discovered a proof of this kind while he was still a student. The exercises that follow pick out the main steps.

The 2000 year hiatus between statement of the problem and its negative solution is even more surprising in view of the progress made by Euclid in Book

X of the *Elements*. As mentioned in Section 3.9, Book X contains many results about numbers of the forms $a+b\sqrt{c}$, $\sqrt{a+b\sqrt{c}}$ and $\sqrt{\sqrt{a}\pm\sqrt{c}}$ where $a,b,c\in\mathbb{Q}$. They include the fact that $a+b\sqrt{c}=a'+b'\sqrt{c}\Rightarrow a=a'$ and $b=b'$ when $\sqrt{c}\notin\mathbb{Q}$ (Proposition 79). This is one of the key steps in showing $\sqrt[3]{2}$ is not constructible. However, Euclid failed to consider arbitrary nesting (probably because his theory was intended for the regular polyhedra) and he did not compare his irrationals with cubic irrationals (for the same reason).

The first to prove inexpressibility of a cubic irrational by particular square roots was Fibonacci [1225], who showed that the roots of $x^3+2x^2+10x=20$ are not any of Euclid's irrationals. He did not prove that the roots are nonconstructible because, like Euclid, he did not consider arbitrary nesting of square roots. It seems that mathematicians were not prepared to consider nested radicals of arbitrary complexity until the 19th century.

The idea of measuring the complexity of a radical expression α by the dimension of the field $\mathbb{Q}(\alpha)$ is due to Dedekind. He developed it in the supplements he wrote for Dirichlet's *Vorlesungen über Zahlentheorie* between 1871 and 1893. (The theorem I have called his "product theorem" appears on p.473 of his work [1893] with the remark that "one easily sees" that the products of basis elements form a basis for an iterated extension.) It is one of the mysteries of mathematical history why the ideas of linear algebra – independence, basis, dimension – were not abstracted earlier than this. They had been implicit in European mathematics for at least two centuries, and in Chinese mathematics for nearly two millennia (the *Nine Chapters of Mathematical Art*, around 300 AD, uses the so-called "Gaussian elimination" method for solving systems of linear equations). It is only when linear algebra is viewed abstractly that the concepts of basis and dimension – and their applications to the nonlinear algebra of algebraic numbers – come to light.

Wantzel [1837] also extended the results of Gauss on regular p-gons. In particular, he showed that the regular p-gon is constructible only when p is a Fermat prime. This filled a gap in Gauss [1801], where the result is claimed but only the (harder) converse is proved. More generally, Wantzel showed that the regular n-gon is constructible only when $n=2^m p_1\cdots p_k$ where $p_1,\ldots,p_k$ are distinct Fermat primes. He did this *without* knowing the irreducibility of $\Phi_n(x)$, instead observing that constructibility of the $p_1^{i_1}\cdots p_k^{i_k}$-gon, for any primes $p_j>2$, implies constructibility of each $p_j^{i_j}$-gon (by selecting suitable vertices). This reduces the irreducibility problem to the special case of $\Phi_q(x)$ for $q=p_j^{i_j}$, which Wantzel was able to handle by a slight modification of Gauss's proof that $\Phi_p(x)$ is irreducible for p prime.

As we have just mentioned, proving constructibility of the regular p-gon when p is a Fermat prime is harder than proving nonconstructibility when p is not. The proof of Gauss [1801], article 365, uses ingenious manipulations of certain sums of roots of the cyclotomic equation $\Phi_p(x)=0$. Some important concepts are involved, but they are obscured by complicated notation. More polished proofs may be found in Hadlock [1978] and Koch [1991], though they labor under the same notational difficulties. To obtain a conceptually clearer proof we need to study not just the dimension of fields but also their *symmetry* – the extent to which different field elements can have the same algebraic behavior.

The next chapter paves the way for this by investigating what it means for two rings or fields to "look the same," or for two elements to have the "same behavior." Clarification of these concepts has many benefits. Among other things, it gives a deeper understanding of the Euler phi function and its role in the study of regular n-gons.

Exercises

5.7.1 If $F_0 = \mathbb{Q}$ and $F_{k+1} = \{a + b\sqrt{c_k} : a, b \in F_k\}$ for some $c_k \in F_k$, show that each F_{k+1} is a field.

5.7.2 Show that if $a, b, c \in F_k$ but $\sqrt{c} \notin F_k$ then $a + b\sqrt{c} = 0 \Leftrightarrow a = b = 0$.

5.7.3 Suppose $\sqrt[3]{2} = a + b\sqrt{c}$ where $a, b, c \in F_k$, but that $\sqrt[3]{2} \notin F_k$. (We know $\sqrt[3]{2} \notin F_0 = \mathbb{Q}$.) Cube both sides and deduce that

$$2 = a^3 + 3ab^2c \quad \text{and} \quad 0 = 3a^2b + b^3c.$$

5.7.4 Deduce from Exercise 5.7.3 that also $\sqrt[3]{2} = a - b\sqrt{c}$, which is a contradiction.

5.7.5 Show that Fibonacci's equation $x^3 + 2x^2 + 10x = 20$ has no rational root and hence that its roots are not constructible.

6 Isomorphisms

6.1 Ring and Field Isomorphisms

In Section 4.5 we introduced the concept of isomorphism to formalise the sense in which $\mathbb{Q}[x]/(x^2-2)\mathbb{Q}[x]$ and $\mathbb{Q}(\sqrt{2})$ are the "same." It is now time to investigate the concept of isomorphism more widely, and we shall begin by extending the definition of isomorphism to rings.

Rings R, R' are said to be *isomorphic* (from the Greek for "same form") if there is a one-to-one function σ from R onto R' which preserves $+$ and $\times$, that is

$$\sigma(a+b) = \sigma(a) + \sigma(b) \text{ for all } a, b \in R,$$
$$\sigma(ab) = \sigma(a)\sigma(b) \text{ for all } a, b \in R.$$

Such a function σ is called an *isomorphism*. It follows from the preservation of $+$ and $\times$ that an isomorphism preserves all ring characteristics, and if R is a field it also preserves the field characteristics. In fact, we have the following theorem.

Theorem. *If σ maps ring R onto ring R' and preserves $+$ and $\times$, then σ also preserves $-$, 0 and 1. If R is a field then σ also preserves $\div$.*

Proof. For any $a \in R$ we have the following implications, which yield preservation of 0 and $\sigma(a-b) = \sigma(a) - \sigma(b)$:

$$\begin{aligned}
&a = a + 0\\
&\Rightarrow \sigma(a) = \sigma(a+0) = \sigma(a) + \sigma(0) \text{ by preservation of } +\\
&\Rightarrow 0 = \sigma(0), \text{ subtracting } \sigma(a) \text{ from both sides}\\
&\Rightarrow 0 = \sigma(a + (-a)) = \sigma(a) + \sigma(-a) \text{ by preservation of } +\\
&\Rightarrow \sigma(-a) = -\sigma(a)\\
&\Rightarrow \sigma(b-a) = \sigma(b + (-a)) = \sigma(b) + \sigma(-a) = \sigma(b) - \sigma(a).
\end{aligned}$$

Since σ is onto R' and we assume each ring has a 1 (compare with Section 2.2), there is some $a \in R$ such that $\sigma(a) = 1$. Then we get preservation of 1 as follows:

$$1 = \sigma(a) = \sigma(a \times 1) = \sigma(a) \times \sigma(1) = 1 \times \sigma(1) = \sigma(1).$$

Finally, if R is a field then there is a b^{-1} for each $b \neq 0$ and

$$\begin{aligned}
&1 = \sigma(bb^{-1}) = \sigma(b)\sigma(b^{-1}) \text{ by preservation of } \times\\
&\Rightarrow \sigma(b^{-1}) = \sigma(b)^{-1}\\
&\Rightarrow \sigma(a/b) = \sigma(ab^{-1}) = \sigma(a)\sigma(b^{-1}) = \sigma(a)\sigma(b)^{-1} = \sigma(a)/\sigma(b).
\end{aligned}$$

□

Corollary. *If $\mathbb{Z} \subseteq R, R'$ then $\sigma(n) = n$ for each $n \in \mathbb{Z}$. If R is also a field then $\sigma(r) = r$ for each $r \in \mathbb{Q}$.*

Proof. Considering natural numbers first, we have:

$$\begin{aligned} n \in \mathbb{N} \Rightarrow n &= 1 + 1 + \cdots + 1 \\ &\Rightarrow \sigma(n) = \sigma(1) + \sigma(1) + \cdots + \sigma(1) = 1 + 1 + \cdots + 1 = n \\ &\Rightarrow \sigma(-n) = -n \text{ by preservation of } - \end{aligned}$$

Thus $n \in \mathbb{Z} \Rightarrow \sigma(n) = n$. Finally, if R is a field then $m/n \in R$ for each $m, n \in \mathbb{Z}$ and we have

$$\sigma(m/n) = \sigma(m)/\sigma(n) = m/n. \qquad \square$$

The corollary shows that the only isomorphism of $\mathbb{Q}$ into $\mathbb{Q}$ is the identity. Thus to find interesting examples of isomorphisms the nearest place to look is $\mathbb{Q}(\alpha)$, where α is an irrational algebraic number.

Example. The function $\sigma : \mathbb{Q}(\sqrt{2}) \to \mathbb{Q}(\sqrt{2})$ defined by $\sigma(a + b\sqrt{2}) = a - b\sqrt{2}$ when $a, b \in \mathbb{Q}$ is an isomorphism.

The fact that σ is one-to-one (and indeed well defined) follows from unique representation of elements of $\mathbb{Q}(\sqrt{2})$ in the form $a + b\sqrt{2}$, which in turn follows from the irrationality of $\sqrt{2}$ (compare with Section 4.5). Since it is likewise true that each element of $\mathbb{Q}(\sqrt{2})$ is expressible in the form $a - b\sqrt{2}$ with $a, b \in \mathbb{Q}$, σ is onto $\mathbb{Q}(\sqrt{2})$. Thus it remains to show that σ preserves $+$ and $\times$. This can be checked by direct calculation: σ preserves $+$ because

$$\begin{aligned} \sigma((a_1 + b_1\sqrt{2}) + (a_2 + b_2\sqrt{2})) &= \sigma(a_1 + a_2 + (b_1 + b_2)\sqrt{2}) \\ &= (a_1 + a_2) - (b_1 + b_2)\sqrt{2} \\ &= (a_1 - b_1\sqrt{2}) + (a_2 - b_2\sqrt{2}) \\ &= \sigma(a_1 + b_1\sqrt{2}) + \sigma(a_2 + b_2\sqrt{2}), \end{aligned}$$

and σ preserves $\times$ because

$$\begin{aligned} \sigma((a_1 + b_1\sqrt{2})(a_2 + b_2\sqrt{2})) &= \sigma(a_1a_2 + 2b_1b_2 + (a_1b_2 + b_1a_2)\sqrt{2}) \\ &= (a_1a_2 + 2b_1b_2) - (a_1b_2 + b_1a_2)\sqrt{2} \\ &= (a_1 - b_1\sqrt{2})(a_2 - b_2\sqrt{2}) \\ &= \sigma(a_1 + b_1\sqrt{2})\sigma(a_2 + b_2\sqrt{2}). \end{aligned}$$

An alternative to this computational proof is to use the isomorphism

$$\mathbb{Q}(\sqrt{2}) \to \mathbb{Q}[x]/(x^2 - 2)\mathbb{Q}[x]$$

found in Section 4.5, which sends $\sqrt{2}$ to (the congruence class of) x. Recall that we actually showed $\mathbb{Q}[x]/(x^2 - 2)\mathbb{Q}[x]$ to be isomorphic to $\mathbb{Q}(\alpha)$ for *any* root α of $x^2 - 2$, hence we also have an isomorphism

$$\mathbb{Q}[x]/(x^2 - 2)\mathbb{Q}[x] \to \mathbb{Q}(\sqrt{2})$$

sending x to $-\sqrt{2}$. Combining the two,

$$\mathbb{Q}(\sqrt{2}) \to \mathbb{Q}[x]/(x^2 - 2)\mathbb{Q}[x] \to \mathbb{Q}(\sqrt{2})$$

where

$$\sqrt{2} \mapsto x \mapsto -\sqrt{2},$$

we get an isomorphism $\sigma : \mathbb{Q}(\sqrt{2}) \to \mathbb{Q}(\sqrt{2})$ with $\sigma(\sqrt{2}) = -\sqrt{2}$. It follows that

$$\begin{aligned}\sigma(a+b\sqrt{2}) &= \sigma(a)+\sigma(b)\sigma(\sqrt{2}) \text{ by preservation of } +, \times \\ &= a + b\sigma(\sqrt{2}) \text{ by preservation of } \mathbb{Q} \\ &= a - b\sqrt{2},\end{aligned}$$

so this is the same as the isomorphism given above. □

The idea of using $\mathbb{Q}[x]/(x^2-2)\mathbb{Q}[x]$ as an abstract bridge connecting $\mathbb{Q}(\sqrt{2})$ to $\mathbb{Q}(\sqrt{2})$ in a nontrivial way easily generalises to a field $F(\alpha)$. We shall prove a general theorem to this effect in the next section, and use it to construct more examples of nontrivial isomorphisms.

Exercises

6.1.1 Give an example of a function ψ mapping a ring R onto a ring R' which preserves $+$ and $\times$ (a *homomorphism*, see Section 6.8) but which is not one-to-one.

6.1.2 Show that any homomorphism ψ of one field onto another is an isomorphism. (Hint: It suffices to show that ψ cannot send a nonzero element to 0.)

6.1.3 Show that the function $\sigma(x) = x^p$ is an isomorphism of the finite field $\mathbb{Z}/p\mathbb{Z}$ onto itself.

6.1.4 Is $\mathbb{Q}(\sqrt{2})$ isomorphic to $\mathbb{Q}(\sqrt{3})$?

6.2 Isomorphisms of $\mathbb{Q}(\alpha)$ and $F(\alpha)$

The isomorphism of $\mathbb{Q}(\sqrt{2})$ sending $\sqrt{2}$ to $-\sqrt{2}$ shows that the two roots of x^2-2 have the "same algebraic behavior" in a precise sense. The following theorem shows that any two roots of the same irreducible polynomial in $\mathbb{Q}[x]$ have the same behavior, in the sense that they correspond under an isomorphism of fields. Conversely, algebraic numbers that correspond under an isomorphism are roots of the same irreducible polynomial.

Conjugation Theorem. *If α, β are algebraic numbers then α, β satisfy the same irreducible $h(x) \in \mathbb{Q}[x]$ $\Leftrightarrow$ there is an isomorphism $\sigma : \mathbb{Q}(\alpha) \to \mathbb{Q}(\beta)$ with $\sigma(\alpha) = \beta$.*

Proof. ($\Rightarrow$) If α, β are roots of the same irreducible $h(x) \in \mathbb{Q}[x]$ then Theorem 4.5 gives isomorphisms

$$\mathbb{Q}(\alpha) \to \mathbb{Q}[x]/h(x)\mathbb{Q}[x] \to \mathbb{Q}(\beta)$$

which send

$$\alpha \mapsto x \mapsto \beta.$$

The composite $\sigma : \mathbb{Q}(\alpha) \to \mathbb{Q}(\beta)$ of these isomorphisms is therefore an isomorphism with $\sigma(\alpha) = \beta$.

($\Leftarrow$) If $\sigma : \mathbb{Q}(\alpha) \to \mathbb{Q}(\beta)$ is an isomorphism with $\sigma(\alpha) = \beta$, let $h(x) \in \mathbb{Q}[x]$ be the irreducible monic polynomial satisfied by α. Applying σ to the equation $0 = h(\alpha)$ we get

$$0 = \sigma(0) = \sigma h(\alpha) = h(\sigma(\alpha))$$

since σ preserves $+, \times$ and $\mathbb{Q}$. Namely, if

$$h(\alpha) = a_m\alpha^m + \cdots + a_1\alpha + a_0 \quad \text{with } a_0, \ldots, a_m \in \mathbb{Q}$$

then

$$\begin{aligned} \sigma h(\alpha) &= \sigma(a_m)(\sigma(\alpha))^m + \cdots + \sigma(a_1)\sigma(\alpha) + \sigma(a_0) \\ &= a_m(\sigma(\alpha))^m + \cdots + a_1\sigma(\alpha) + a_0 \\ &= h(\sigma(\alpha)). \end{aligned}$$

Thus $\sigma(\alpha) = \beta$ is also a root of $h(x)$. □

This theorem gives many isomorphisms for which the preservation of $+$ and $\times$ would be laborious to check directly.

Example 1. $\sigma : \mathbb{Q}(\sqrt[3]{2}) \to \mathbb{Q}(\zeta_3\sqrt[3]{2})$ with $\sigma(\sqrt[3]{2}) = \zeta_3\sqrt[3]{2}$ is an isomorphism.

This is because $\sqrt[3]{2}, \zeta_3\sqrt[3]{2}$ are both roots of the polynomial $x^3 - 2$, which is irreducible over $\mathbb{Q}$ (for example by Eisenstein's criterion). □

The opposite direction of the theorem gives

Example 2. The only isomorphism $\sigma : \mathbb{Q}(\sqrt[3]{2}) \to \mathbb{Q}(\sqrt[3]{2})$ is the identity.

This is because the only roots of $x^3 - 2$ are $\sqrt[3]{2}, \zeta_3\sqrt[3]{2}$ and $\zeta_3^2\sqrt[3]{2}$, and the latter two are not real, hence not members of $\mathbb{Q}(\sqrt[3]{2})$. □

I call the theorem the conjugation theorem because the roots $\alpha_1, \alpha_2, \ldots$ of an irreducible $h(x) \in \mathbb{Q}[x]$ are called *conjugates* of each other. This generalises the use of the word "conjugate" in "complex conjugate" (complex conjugation $a+bi \mapsto a-bi$ is an isomorphism $\mathbb{C} \to \mathbb{C}$) and "conjugate surds" (when $a, b, c \in \mathbb{Q}$ and $\sqrt{c}$ is irrational, $a + b\sqrt{c} \mapsto a - b\sqrt{c}$ is an isomorphism $\mathbb{Q}(\sqrt{c}) \to \mathbb{Q}(\sqrt{c})$). Thus the theorem may be interpreted as saying that the isomorphisms of $\mathbb{Q}(\alpha)$ onto other number fields are all "conjugations."

There is a generalisation to fields $F(\alpha)$, whose statement and proof become clear once one realises that the only property of $\mathbb{Q}$ used in the proof of the conjugation theorem is that $\sigma(a) = a$ for all $a \in \mathbb{Q}$.

General Conjugation Theorem. *If α, β are algebraic over F then α, β satisfy the same irreducible $h(x) \in F[x] \Leftrightarrow$ there is an isomorphism $\sigma : F(\alpha) \to F(\beta)$ with $\sigma(\alpha) = \beta$ and $\sigma(a) = a$ for all $a \in F$.*

Proof. ($\Rightarrow$) Using the isomorphisms from Section 4.5

$$F(\alpha) \to F[x]/h(x)F[x] \to F(\beta)$$

which send

$$\alpha \mapsto x \mapsto \beta$$

(and also fix each $a \in F$), we get a composite isomorphism $\sigma : F(\alpha) \to F(\beta)$ with the required properties.

($\Leftarrow$) Given an isomorphism $\sigma : F(\alpha) \to F(\beta)$ with $\sigma(\alpha) = \beta$ and $\sigma(a) = a$ for all $a \in F$, we apply σ to the equation $0 = h(\alpha)$ and get $0 = h(\sigma(\alpha)) = h(\beta)$, using the fact that $\sigma(a) = a$ for the coefficients a of $h(\alpha)$. □

The general conjugation theorem enables us to find isomorphisms of complicated fields $\mathbb{Q}(\alpha, \beta, \ldots)$ by viewing them as $F(\alpha)$, where F is simpler. Bear in mind that an isomorphism σ of $\mathbb{Q}(\alpha, \beta, \ldots)$ is determined by the values of $\sigma(\alpha), \sigma(\beta), \ldots$, since for any rational function $f(\alpha, \beta, \ldots) \in \mathbb{Q}(\alpha, \beta, \ldots)$ we have

$$\sigma f(\alpha, \beta, \ldots) = f(\sigma(\alpha), \sigma(\beta), \ldots)$$

by the preservation of $+, -, \times, \div$ and coefficients $a \in \mathbb{Q}$.

Example 3. There is an isomorphism $\sigma : \mathbb{Q}(\sqrt[3]{2}, \zeta_3) \to \mathbb{Q}(\sqrt[3]{2}, \zeta_3)$ with $\sigma(\sqrt[3]{2}) = \sqrt[3]{2}$, $\sigma(\zeta_3) = \zeta_3^2$.

View $\mathbb{Q}(\sqrt[3]{2}, \zeta_3)$ as $F(\zeta_3)$, where $F = \mathbb{Q}(\sqrt[3]{2})$. Now ζ_3, ζ_3^2 (the complex cube roots of unity) are the roots of $x^2 + x + 1$, which is irreducible over $F = \mathbb{Q}(\sqrt[3]{2})$ because F is real and the roots of $x^2 + x + 1$ are not. Thus it follows from the general conjugation theorem that there is an isomorphism σ of $F(\zeta_3) = \mathbb{Q}(\sqrt[3]{2}, \zeta_3)$ with $\sigma(\zeta_3) = \zeta_3^2$ and $\sigma(a) = a$ for all $a \in F$, in particular $\sigma(\sqrt[3]{2}) = \sqrt[3]{2}$. □

Example 4. There is an isomorphism $\sigma : \mathbb{Q}(\sqrt[3]{2}, \zeta_3) \to \mathbb{Q}(\sqrt[3]{2}, \zeta_3)$ with $\sigma(\sqrt[3]{2}) = \zeta_3\sqrt[3]{2}$, $\sigma(\zeta_3) = \zeta_3$.

View $\mathbb{Q}(\sqrt[3]{2}, \zeta_3)$ as $F(\sqrt[3]{2})$, where $F = \mathbb{Q}(\zeta_3)$. Since $\sqrt[3]{2}$, $\zeta_3\sqrt[3]{2}$ are roots of $x^3 - 2$, it suffices to show that $x^3 - 2$ is irreducible over $F = \mathbb{Q}(\zeta_3)$. We know that $x^3 - 2$ is irreducible over $\mathbb{Q}$ (for example, by Eisenstein's criterion), which means that its roots all have degree 3 over $\mathbb{Q}$ (Section 5.2). If $x^3 - 2$ is not irreducible over $\mathbb{Q}(\zeta_3)$ it has at least one linear factor ($x -$ something) in $\mathbb{Q}(\zeta_3)$, which means that at least one of its roots is in $\mathbb{Q}(\zeta_3)$. This is impossible because ζ_3 is of degree 2 over $\mathbb{Q}$ (as a root of $x^2 + x + 1$), hence all elements of $\mathbb{Q}(\zeta_3)$ have degree ≤ 2 over $\mathbb{Q}$ by Section 5.4. □

Exercises

6.2.1 Deduce from the irreducibility of $x^4 - 2$ over $\mathbb{Q}(i)$ (Section 5.5) that $\sigma(\sqrt[4]{2}) = i\sqrt[4]{2}$, $\sigma(i) = i$ defines an isomorphism of $\mathbb{Q}(\sqrt[4]{2}, i)$. Also find a nontrivial isomorphism that fixes $\sqrt[4]{2}$.

6.2.2 Find the conjugates of $\alpha = \sqrt{2} + \sqrt{3}$ by finding all roots of the minimal polynomial $x^4 - 10x^2 + 1$ for α found in Exercise 5.5.3.

6.3 Extending Fields and Isomorphisms

The general conjugation theorem of Section 6.2 can be complemented by a theorem about isomorphisms of $F(\alpha)$ which do *not* fix all the elements of F, but in fact extend arbitrary isomorphisms of F. Such isomorphisms may be found by observing how an isomorphism τ of F propagates through the construction of $F(\alpha)$ as $F[x]/h(x)F[x]$ in Section 4.5. The first step is the following lemma on the extension of τ from F to $F[x]$, which we shall also use in Section 6.5*.

Lemma. *An isomorphism* $\tau : F \to \tau F$ *of fields extends to an isomorphism* $\tau : F[x] \to \tau F[x]$ *by setting* $\tau(x) = x$.

Proof. If $f(x) = a_0 + a_1x + \cdots + a_mx^m \in F[x]$ we define

$$\tau f(x) = \tau(a_0) + \tau(a_1)x + \cdots + \tau(a_m)x^m.$$

Thus if

$$g(x) = b_0 + b_1x + \cdots + b_nx^n$$

we have

$$\tau g(x) = \tau(b_0) + \tau(b_1)x + \cdots + \tau(b_n)x^n,$$

and it follows immediately from the fact that $\tau(a+b) = \tau(a)+\tau(b)$ for $a, b \in F$ that

$$\tau(f(x) + g(x)) = \tau f(x) + \tau g(x).$$

Similarly (but with a bit more calculation, and using the additional fact that $\tau(ab) = \tau(a)\tau(b)$ for $a, b \in F$), we find

$$\tau(f(x)g(x)) = \tau f(x)\tau g(x).$$

Finally, since τ is one-to-one on F it follows that $\tau f(x), \tau g(x)$ have the same coefficients only when $f(x), g(x)$ have. Thus τ is also one-to-one on $F[x]$, and hence an isomorphism. □

Thus the natural extension of $\tau : F \to \tau F$ to a map $F[x] \to \tau F[x]$ is an isomorphism. The theorem is obtained by showing that the latter isomorphism determines a one-to-one correspondence between the congruence classes of $F[x]$ mod $h(x)$ and the congruence classes of $\tau F[x]$ mod $\tau h(x)$, and hence an isomorphism of $F(\alpha) = F[x]/h(x)F[x]$ onto $\tau F[x]/\tau h(x)\tau F[x]$.

Isomorphism Extension Theorem. *If* α *is algebraic over a field* F *then any isomorphism of* F *extends to an isomorphism of* $F(\alpha)$.

Proof. If $\alpha \in F$ there is nothing to prove, so suppose $\alpha \notin F$ and that α satisfies an irreducible $h(x) \in F[x]$.

If $\tau : F \to \tau F$ is an isomorphism, we consider the extended isomorphism $\tau : F[x] \to \tau F[x]$ given by the lemma. Since τ preserves $+, -$ and $\times$ it follows that

$$f(x) - g(x) = q(x)h(x) \ \Rightarrow\ \tau f(x) - \tau g(x) = \tau q(x)\tau h(x),$$

and the converse also holds since τ is one-to-one. Thus

$$f(x) \equiv g(x) \pmod{h(x)} \Leftrightarrow \tau f(x) \equiv \tau g(x) \pmod{\tau h(x)}$$

and hence we have a one-to-one correspondence $\tilde{\tau}$ between congruence classes defined by

$$\tilde{\tau}(h(x)F[x] + f(x)) = \tau h(x)\tau F[x] + \tau f(x).$$

Since τ preserves $+$ and $\times$ of representatives $f(x), \tilde{\tau}$ preserves $+$ and $\times$ of their congruence classes, and hence it is an isomorphism of $F[x]/h(x)F[x]$ onto $\tau F[x]/\tau h(x)\tau F[x]$.

Moreover, for congruence classes of elements $a \in F$ we have

$$\tilde{\tau}(h(x)F[x] + a) = \tau h(x)F[x] + \tau a.$$

That is, $\tilde{\tau}$ sends the class of $a \in F$ in $F[x]/h(x)F[x] = F(\alpha)$ to the class of τa, and hence it is an extension to $F(\alpha)$ of the isomorphism τ of F. □

Example 1. Isomorphisms of $\mathbb{Q}(\sqrt{2}, \sqrt{3})$.

We can view $\mathbb{Q}(\sqrt{2}, \sqrt{3})$ as an extension of either $\mathbb{Q}(\sqrt{2})$ or $\mathbb{Q}(\sqrt{3})$. Viewing it first as $F(\sqrt{3})$, where $F = \mathbb{Q}(\sqrt{2})$, we look at what happens when we extend the isomorphism τ_1 of $\mathbb{Q}(\sqrt{2})$ such that $\tau_1(\sqrt{2}) = -\sqrt{2}$. The adjoined element $\sqrt{3}$ does not belong to $\mathbb{Q}(\sqrt{2})$, otherwise

$$\sqrt{3} = a + b\sqrt{2} \text{ for some } a, b \in \mathbb{Q}(\sqrt{2}),$$

and then

$$3 = (a + b\sqrt{2})^2 = a^2 + 2b^2 + 2ab\sqrt{2}$$

which yields the contradiction

$$\sqrt{2} = \frac{3 - a^2 - 2b^2}{2ab} \in \mathbb{Q}.$$

Thus we can extend τ_1 to an isomorphism $\tilde{\tau}_1$ of $F(\sqrt{3}) = F[x]/(x^2-3)F[x]$ and by construction $\tilde{\tau}_1(\sqrt{3}) = \sqrt{3}$ because $\tilde{\tau}_1(x) = x$ and $x = \sqrt{3}$ or $-\sqrt{3}$. Thus $\tilde{\tau}_1$ is an isomorphism of $\mathbb{Q}(\sqrt{2}, \sqrt{3})$ such that $\tilde{\tau}_1(\sqrt{2}) = -\sqrt{2}$, $\tilde{\tau}_1(\sqrt{3}) = \sqrt{3}$.

The analogous construction beginning with the isomorphism τ_2 of $\mathbb{Q}(\sqrt{3})$ such that $\tau_2(\sqrt{3}) = -\sqrt{3}$ yields an isomorphism $\tilde{\tau}_2$ of $\mathbb{Q}(\sqrt{2}, \sqrt{3})$ such that $\tilde{\tau}_2(\sqrt{2}) = \sqrt{2}$, $\tilde{\tau}_2(\sqrt{3}) = -\sqrt{3}$. We get a third isomorphism as the composite $\tilde{\tau}_1\tilde{\tau}_2$, which changes the signs of both $\sqrt{2}$ and $\sqrt{3}$. And fourthly, of course, there is the identity isomorphism which leaves $\sqrt{2}$ and $\sqrt{3}$ unaltered.

Now we can check that any isomorphism of $\mathbb{Q}(\sqrt{2}, \sqrt{3})$ onto itself is one of the four just found. Any isomorphism σ of $\mathbb{Q}(\sqrt{2}, \sqrt{3})$ is determined by the values of $\sigma(\sqrt{2})$ and $\sigma(\sqrt{2})$, as we observed more generally for $\mathbb{Q}(\alpha, \beta, \ldots)$ in Section 6.2. The value $\sigma(\sqrt{2})$ must satisfy

$$(\sigma(\sqrt{2}))^2 = \sigma(\sqrt{2})\sigma(\sqrt{2}) = \sigma(\sqrt{2}\sqrt{2}) = \sigma(2) = 2,$$

hence $\sigma(\sqrt{2}) = \pm\sqrt{2}$, and similarly $\sigma(\sqrt{3}) = \pm\sqrt{3}$. This gives four possibilities for σ, and they are exactly the possibilities realised by the isomorphisms already found. □

In Example 1 we can get away with the naive interpretation of x (as $\sqrt{3}$ or $-\sqrt{3}$) because the polynomial $h(x) = x^2 - 3$ satisfied by x is unchanged by the isomorphism τ of $F[x]$. In $F[x]/h(x)F[x]$, x is a root of $x^2 - 3$, and in $\tau F[x]/\tau h(x)\tau F[x]$ it is a root of $\tau(x^2 - 3)$, which also equals $x^2 - 3$. In general, the numerical interpretation of x will be different in $\tau F[x]/\tau h(x)\tau F[x]$ because $\tau h(x)$ will be a different polynomial. The point is brought out by the following example, suggested to me by Angelo di Pasquale.

Example 2. An isomorphism of $\mathbb{Q}(\sqrt[4]{2})$.

We can view $\mathbb{Q}(\sqrt[4]{2})$ as $F(\sqrt[4]{2})$, where $F = \mathbb{Q}(\sqrt{2})$, and construct an isomorphism $\tilde{\tau}$ extending the isomorphism $\tau : F \to F$ defined by $\tau(\sqrt{2}) = -\sqrt{2}$. Again $\tilde{\tau}(x) = x$, however, the numerical interpretation of x changes between F and τF. Why? Because the irreducible polynomial satisfied by $\sqrt[4]{2}$, $x^2 - \sqrt{2} \in F[x]$, changes to $\tau(x^2 - \sqrt{2}) = x^2 + \sqrt{2}$. Hence in $\tau F[x]/\tau(x^2 - \sqrt{2})\tau F[x]$, x satisfies $\tau(x^2 - \sqrt{2}) = x^2 + \sqrt{2}$, which is *not* satisfied by $x = \sqrt[4]{2}$. We have to interpret $\tilde{\tau}(x) = \tau(\sqrt[4]{2})$ as a root of $x^2 + \sqrt{2}$, say $\tilde{\tau}(\sqrt[4]{2}) = i\sqrt[4]{2}$, and then $\tilde{\tau}(\sqrt{2}) = (\tilde{\tau}(\sqrt[4]{2}))^2 = -\sqrt{2}$ as required. □

Remark. The theorem is still true when α is not algebraic over F, and in fact a bit simpler. In this case $F(\alpha)$ is isomorphic to the field $F(x)$ of rational functions in x with coefficients in F (compare with Section 5.1), so it is only necessary to generalise the lemma from $F[x]$ to $F(x)$. We do not carry this out because we do not need the result. However, we shall be looking at isomorphisms of rational function fields, for other reasons, in Section 6.5*. The theorem itself plays a brief but important role in Chapters 8 and 9.

Exercises

6.3.1 Give another proof that the mappings $\tilde{\tau}_1, \tilde{\tau}_2$ of $\mathbb{Q}(\sqrt{2}, \sqrt{3})$ are isomorphisms by using the general conjugation theorem of Section 6.2.

6.3.2 Give yet another proof, by viewing $\mathbb{Q}(\sqrt{2}, \sqrt{3})$ as $\mathbb{Q}(\sqrt{2} + \sqrt{3})$ and interpreting $\tilde{\tau}_1$ and $\tilde{\tau}_2$ as conjugations of $\sqrt{2} + \sqrt{3}$ (using Exercise 6.2.2).

6.3.3 Use extensions to give another proof that the mappings in Examples 3 and 4 of Section 6.2 are isomorphisms of $\mathbb{Q}(\sqrt[3]{2}, \zeta_3)$.

6.3.4 Show that six combinations of values of $\sigma(\sqrt[3]{2})$ and $\sigma(\zeta_3)$ are possible for an isomorphism of $\mathbb{Q}(\sqrt[3]{2}, \zeta_3)$ onto itself, and that all six actually occur for composites of the isomorphisms found in Section 6.2.

6.3.5 Show that any isomorphisms of $\mathbb{Q}(\cos \frac{2\pi}{n})$ are isomorphisms of $\mathbb{Q}(\zeta_n)$.

6.4 Automorphisms and Groups

An obvious property of isomorphisms, already used in the last theorem and elsewhere, is that the inverse σ^{-1} of an isomorphism σ is also an isomorphism, as is the composite $\sigma_1\sigma_2$ of isomorphisms σ_1 and σ_2. In this and the following section we shall frequently construct new isomorphisms from old by composition and inversion. In particular, we shall use the idea to construct *automorphisms* of a ring R, that is, isomorphisms of R onto itself.

Automorphisms are the most manageable isomorphisms, because not only are the composite and inverse of automorphisms of R again automorphisms of R, there is also an *identity* automorphism of R, namely the identity function $\mathbf{1}$ defined by $\mathbf{1}(r) = r$ for all $r \in R$. The fundamental properties of automorphisms under the operation of composition are

$$\sigma(\sigma'\sigma'') = (\sigma\sigma')\sigma'' \qquad (associativity)$$

which is true for composition of *any* functions,

$$\sigma\mathbf{1} = \mathbf{1}\sigma = \sigma \qquad (identity)$$

which is the characteristic property of the identity function, and

$$\sigma\sigma^{-1} = \sigma^{-1}\sigma = \mathbf{1} \qquad (inverse)$$

which is the characteristic property of inverse functions.

These three properties define a *group of one-to-one functions* under the operation of composition. In general, a *group* is any set of objects under an operation with the associative, identity and inverse properties. However, there is a theorem of Cayley (Section 7.2) which shows that any group is essentially the same as a group of one-to-one functions under composition. Thus there is no loss of generality in assuming the group operation to be function composition. We shall in fact be interested only in groups of automorphisms of fields, but even these turn out to be quite general. They include all the finite groups, and some particularly interesting examples will be exhibited in Chapter 7.

Exercise

6.4.1 Without assuming that group elements are functions, deduce from the three properties of a group that

$$\sigma\sigma_1 = \sigma\sigma_2 \;\Rightarrow\; \sigma_1 = \sigma_2 \qquad (cancellation)$$

$$\sigma\sigma_1 = \sigma_1 \;\Rightarrow\; \sigma = \mathbf{1} \qquad (uniqueness\ of\ identity)$$

$$\sigma\sigma_1 = \mathbf{1} \;\Rightarrow\; \sigma = \sigma_1^{-1} \qquad (uniqueness\ of\ inverse)$$

6.5* Function Fields and Symmetric Functions

Just as the ring $\mathbb{Z}$ of integers may be extended to the field $\mathbb{Q}$ of rational numbers by forming quotients of integers, each ring $F[x]$ of polynomials may be extended to a field $F(x)$ of *rational functions* by forming quotients of polynomials. We can generalise further, to any finite number of indeterminates, by first forming the ring $F[x_1, \ldots, x_n] = F[x_1] \ldots [x_n]$ of polynomials in $x_1, \ldots, x_n$, and then the field $F(x_1, \ldots, x_n)$ of quotients of these polynomials. Here F can be any field.

In Chapter 8 we shall be interested in the field $\mathbb{Q}(x_1, \ldots, x_n)$ and its subfield $\mathbb{Q}(a_0, a_1, \ldots, a_{n-1})$, where $a_0, a_1, \ldots, a_{n-1}$ are polynomials in $x_1, \ldots, x_n$ defined by

$$(x - x_1) \cdots (x - x_n) = x^n + a_{n-1}x^{n-1} + \cdots + a_1 x + a_0.$$

We shall see that the relationship between these fields is the key to deciding whether the general n^{th} degree equation

$$x^n + a_{n-1}x^{n-1} + \cdots + a_1 x + a_0 = 0$$

is solvable by radicals. The present section gives an optional preview of this relationship.

The first thing to observe is that *each permutation σ of $x_1, \ldots, x_n$ extends to an automorphism of* $\mathbb{Q}(x_1, \ldots, x_n)$, namely the mapping σ of $\mathbb{Q}(x_1, \ldots, x_n)$ that sends each rational function $f(x_1, \ldots, x_n)$ to $f(\sigma(x_1), \ldots, \sigma(x_n))$. This mapping obviously preserves $+$ and $\times$, and it is one-to-one because it is inverted by the mapping σ^{-1} that sends each function $g(x_1, \ldots, x_n)$ to $g(\sigma^{-1}(x_1), \ldots, \sigma^{-1}(x_n))$.

A key property of these permutation automorphisms σ is that they *fix* the polynomials $a_0, a_1, \ldots, a_{n-1}$, that is, $\sigma(a_i) = a_i$ for each a_i. This is because a_i is the coefficient of x^i in the expansion of $(x - x_1) \cdots (x - x_n)$, which is the same as $(x - \sigma(x_1)) \cdots (x - \sigma(x_n))$ for any permutation σ. More surprisingly, the converse is also true: any automorphism of $\mathbb{Q}(x_1, \ldots, x_n)$ fixing $a_0, a_1, \ldots, a_{n-1}$ is the extension of a permutation σ. The proof is a simple combination of the technique of extending an isomorphism from F to $F[x]$ (Lemma 6.2) with unique factorisation in $F[x]$ (Theorem 4.3).

Theorem. *An automorphism σ of $\mathbb{Q}(x_1, \ldots, x_n)$ fixes $a_0, a_1, \ldots, a_{n-1} \Leftrightarrow \sigma$ is the extension of a permutation of $x_1, \ldots, x_n$.*

Proof. For any automorphism σ of $\mathbb{Q}(x_1, \ldots, x_n)$ consider its extension (also called σ) to $\mathbb{Q}(x_1, \ldots, x_n)[x]$ such that $\sigma(x) = x$. Then

$$\begin{aligned}
\sigma \text{ fixes } a_0, \ldots, a_{n-1} &\Leftrightarrow \sigma \text{ fixes } x^n + a_{n-1}x^{n-1} + \cdots + a_1 x + a_0 \\
&\Leftrightarrow \sigma \text{ fixes } (x - x_1) \cdots (x - x_n) \\
&\Leftrightarrow (x - \sigma(x_1)) \cdots (x - \sigma(x_n)) = (x - x_1) \cdots (x - x_n) \\
&\Leftrightarrow \sigma \text{ permutes } x_1, \cdots, x_n,
\end{aligned}$$

by the unique factorisation of polynomials in $\mathbb{Q}(x_1, \ldots, x_n)[x]$. □

Functions fixed by permutations of $x_1, \ldots, x_n$ are called *symmetric*, and the particular polynomials $a_0, a_1, \ldots, a_{n-1}$ are called the *elementary symmetric functions.* The symmetry of a_0 and a_{n-1} is particularly plain because

$$a_0 = (-1)^n x_1 \cdots x_n \text{ and } a_{n-1} = -(x_1 + \cdots + x_n).$$

A remarkable theorem, apparently due to Newton [1707], states that every symmetric function of $x_1, \ldots, x_n$ is in $\mathbb{Q}(a_0, a_1, \ldots, a_{n-1})$. This called the *fundamental theorem of symmetric functions.* In fact, Newton proved the stronger result that any symmetric polynomial is a polynomial in $a_0, \ldots, a_{n-1}$ (see Exercise 6.5.1). A more conceptual proof of the fundamental theorem is obtainable from Galois theory (Exercise 9.3.4).

Exercise

6.5.1 Give a proof by induction that a symmetric polynomial f in $x_1, \ldots, x_n$ is a polynomial in $a_0, \ldots, a_{n-1}$ as follows:

(i) Defining $a_1^0, \ldots, a_{n-1}^0$ by

$$a_i^0(x_1, \ldots, x_{n-1}) = a_i(x_1, \ldots, x_{n-1}, 0),$$

show that the a_i^0 are symmetric in $x_1, \ldots, x_{n-1}$ and, by an induction hypothesis on n, that

$$f^0(x_1, \ldots, x_{n-1}) = f(x_1, \ldots, x_{n-1}, 0)$$

is a polynomial $g(a_1^0, \ldots, a_{n-1}^0)$.

(ii) Show that

$$p(x_1, \ldots, x_n) = f(x_1, \ldots, x_n) - g(a_1, \ldots, a_{n-1})$$

vanishes for $x_n = 0$, hence is divisible by x_n.

(iii) Deduce, by symmetry, that $p(x_1, \ldots, x_n)$ is divisible by all x_i, and hence by a_0.

(iv) Deduce that

$$f(x_1, \ldots, x_n) = g(a_1, \ldots, a_{n-1}) + a_0 h(x_1, \ldots, x_n)$$

where h is symmetric.

(v) Assuming, by induction on degree, that h is a polynomial in $a_0, \ldots, a_{n-1}$, conclude the same for f.

6.6* Cyclotomic Fields

Probably the most important algebraic extensions of $\mathbb{Q}$ are the cyclotomic fields $\mathbb{Q}(\zeta_n)$ where ζ_n is a primitive n^{th} root of unity (Section 3.7). They are intimately involved in the study of nth roots, since the nth roots of any $a \in \mathbb{C}$ are of the form $\sqrt[n]{a}, \zeta_n \sqrt[n]{a}, \ldots, \zeta_n^{n-1} \sqrt[n]{a}$ where $\sqrt[n]{a}$ denotes a particular n^{th} root of a, and the following information is relevant (though not strictly necessary) for our investigation of solution by radicals in Chapters 8 and 9.

Theorem. *The automorphisms of* $\mathbb{Q}(\zeta_n)$ *are the functions* $\sigma_k : \mathbb{Q}(\zeta_n) \to \mathbb{Q}(\zeta_n)$ *such that* $\sigma_k(\zeta_n) = \zeta^k$ *where* $1 \le k \le n$ *and* $\gcd(k, n) = 1$.

Proof. As we know from Section 6.4, any automorphism σ of $\mathbb{Q}(\zeta_n)$ is determined by the value of $\sigma(\zeta_n)$, and

$$(\sigma(\zeta_n))^n = \sigma(\zeta_n^n) = \sigma(1) = 1.$$

Thus $\sigma(\zeta_n)$ is an nth root of unity and therefore

$$\sigma(\zeta_n) = \zeta_n^k \quad \text{where} \quad 1 \le k < n.$$

Also, since $\zeta_n^d \ne 1$ for $1 \le d < n$ we must have $1 \ne (\sigma(\zeta_n))^d = \zeta_n^{kd}$ and hence $kd \not\equiv 0 \pmod{n}$. Thus k is invertible mod n and hence $\gcd(k, n) = 1$ by Theorem 2.7.

Conversely, if $\gcd(k, n) = 1$, then ζ_n^k is a primitive n^{th} root of unity, hence a root of the cyclotomic polynomial $\Phi_n(x)$, by definition of Φ_n (Section 4.8*). Since $\Phi_n(x)$ is irreducible, by Section 4.9*, we have an isomorphism

$$\sigma_k : \mathbb{Q}(\zeta_n) \to \mathbb{Q}(\zeta_n) \quad \text{where} \quad \sigma_k(\zeta_n) = \zeta_n^k$$

by the conjugation theorem of Section 6.2. □

This theorem is essentially equivalent to the irreducibility of $\Phi_n(x)$, since irreducibility is relatively easy to prove from the assumption that $\sigma_k(\zeta_n) = \zeta_n^k$ defines an automorphism of $\mathbb{Q}(\zeta_n)$ when k is relatively prime to n (see Exercises). Note that, in the special case where n is a prime p, the function $\sigma_k(\zeta_p) = \zeta_p^k$ defines an automorphism of $\mathbb{Q}(\zeta_p)$ for $k = 1, 2, \ldots, p-1$.

Note also that the number of automorphisms of $\mathbb{Q}(\zeta_n)$ is the same as its dimension over $\mathbb{Q}$, namely, the degree $\phi(n)$ of $\Phi_n(x)$. This fact is part of the relationship between groups and fields that will unfold in Chapters 7, 8 and 9.

Exercises

6.6.1 Suppose $f(x) \in \mathbb{Q}[x]$ is a nontrivial irreducible factor of $\Phi_n(x)$ such that $f(\zeta_n) = 0$ and suppose that

$$f(x) = (x - \zeta_n)(x - \zeta_n^{i_1}) \ldots (x - \zeta_n^{i_j}) \quad \text{in} \quad \mathbb{Q}(\zeta_n)[x].$$

If an isomorphism σ of $\mathbb{Q}(\zeta_n)$ is extended to $\mathbb{Q}(\zeta_n)[x]$ by setting $\sigma(x) = x$, show that

$$\sigma f(x) = (x - \sigma(\zeta_n))(x - \sigma(\zeta_n^{i_1})) \ldots (x - \sigma(\zeta_n^{i_j}))$$

is just a permutation of the factorisation of $f(x)$.

6.6.2 Deduce from Exercise 6.6.1 that $f(x)$ contains each factor $(x - \zeta_n^k)$ where $\gcd(k, n) = 1$, and hence that $f(x) = \Phi_n(x)$ up to a constant factor, so $\Phi_n(x)$ is irreducible.

6.6.3 Show that for each k relatively prime to n there is an automorphism σ_k of $\mathbb{Q}(\cos\frac{2\pi}{n})$ defined by $\sigma_k(\cos\frac{2\pi}{n}) = \cos\frac{2k\pi}{n}$, and that every automorphism of $\mathbb{Q}(\cos\frac{2\pi}{n})$ is of this form (compare with Exercise 6.3.5).

6.6.4 How many of the automorphisms σ_k of $\mathbb{Q}(\cos\frac{2\pi}{n})$ are actually distinct? Compare this number with the value of $(\mathbb{Q}(\cos\frac{2\pi}{n}) : \mathbb{Q})$ found in Exercise 5.5.4.

6.6.5 Using Exercise 6.6.4 or otherwise, find fields $\mathbb{Q}(\cos\frac{2\pi}{n})$ with automorphism groups of three and five elements.

6.7* The Chinese Remainder Theorem

Most of the ring isomorphisms in this book are actually field isomorphisms, or else simple extensions of field isomorphisms to rings. However, there is one that really is about rings. It comes from reflecting on an ancient discovery known as the Chinese remainder theorem.

In the Chinese *Mathematical Classic of Sun Tzu* (around 300 AD), a method is given for finding a number r, given its remainders $r \bmod m$ and $r \bmod n$ by relatively prime integers m and n. The implication is that, given $r \bmod m$ and $r \bmod n$, the value of $r \bmod mn$ is uniquely determined. This much (or its obvious generalisation to any number of relatively prime integers $m, n, \ldots$) is the classical Chinese remainder theorem. However, not only does the pair $(r \bmod m, r \bmod n)$ determine $r \pmod{mn}$, it turns out that any two integers r and s can be *added* or *multiplied* $\pmod{mn}$ by adding or multiplying their remainder pairs term by term, that is, forming $(r + s \bmod m, r + s \bmod n)$ or $(rs \bmod m, rs \bmod n)$.

This further discovery indicates that the underlying reason for the Chinese remainder theorem is an *isomorphism*, between the ring $\mathbb{Z}/mn\mathbb{Z}$ and the set of pairs $(r \bmod m,\ r \bmod n)$ under term by term addition and multiplication. The latter structure is called the *direct product* of the rings $\mathbb{Z}/m\mathbb{Z}$ and $\mathbb{Z}/n\mathbb{Z}$ and is denoted by $(\mathbb{Z}/m\mathbb{Z}) \times (\mathbb{Z}/n\mathbb{Z})$. In general, if A and B are any two rings their direct product is the set

$$A \times B = \{(a, b) : a \in A, b \in B\},$$

with $+$ and $\times$ defined by

$$(a_1, b_1) + (a_2, b_2) = (a_1 + a_2, b_1 + b_2)$$
$$(a_1, b_1) \times (a_2, b_2) = (a_1 a_2, b_1 b_2).$$

It is easily checked (Exercise 6.7.1) that $A \times B$ is a ring under these operations. Thus the "real" Chinese remainder theorem is the following theorem about isomorphic rings.

Chinese Remainder Theorem. *If* $\gcd(m,n) = 1$ *then the map* $\rho(r) = (r \text{ mod } m, r \text{ mod } n)$ *is an isomorphism of* $\mathbb{Z}/mn\mathbb{Z}$ *onto* $(\mathbb{Z}/m\mathbb{Z}) \times (\mathbb{Z}/n\mathbb{Z})$.

Proof. It is immediate from the definition of ρ and the definition of $+$ and $\times$ in $(\mathbb{Z}/m\mathbb{Z}) \times (\mathbb{Z}/n\mathbb{Z})$ that ρ preserves $+$ and $\times$. To see that ρ is one-to-one suppose that $\rho(r) = \rho(r')$, that is, that

$$(r \text{ mod } m, r \text{ mod } n) = (r' \text{ mod } m, r' \text{ mod } n).$$

It follows from the preservation of $+$ (and hence $-$) that

$$\rho(r - r') = (0, 0),$$

that is, $r - r'$ is divisible by both m and n. Since $\gcd(m,n) = 1$, it follows that mn divides $r - r'$, hence $r = r'$ in $\mathbb{Z}/mn\mathbb{Z}$.

Finally, ρ must be onto $(\mathbb{Z}/m\mathbb{Z}) \times (\mathbb{Z}/n\mathbb{Z})$ since $\mathbb{Z}/mn\mathbb{Z}$ has mn members and so has $(\mathbb{Z}/m\mathbb{Z}) \times (\mathbb{Z}/n\mathbb{Z})$. □

This version of the Chinese remainder theorem also yields the multiplicative property of the Euler phi function, previously proved (with greater difficulty) in Section 2.9*. Quite an impressive bonus, since the classical Chinese remainder theorem said nothing whatever about phi!

Corollary. *If* $\gcd(m,n) = 1$ *then* $\phi(mn) = \phi(m)\phi(n)$.

Proof. Recall from Section 2.8* that $\phi(r)$ is the number of integers $k, 1 \leq k \leq r$, that are relatively prime to r, and also the number of such k with a multiplicative inverse mod r.

Thus $\phi(mn)$ is the number of invertible elements of $\mathbb{Z}/mn\mathbb{Z}$ and hence, by the Chinese remainder theorem, the number of invertible elements of $(\mathbb{Z}/m\mathbb{Z}) \times (\mathbb{Z}/n\mathbb{Z})$. But it follows from the definition of multiplication in $(\mathbb{Z}/m\mathbb{Z}) \times (\mathbb{Z}/n\mathbb{Z})$ that a pair (k,l) is invertible if and only if k is invertible in $\mathbb{Z}/m\mathbb{Z}$ and l is invertible in $\mathbb{Z}/n\mathbb{Z}$. Since $\mathbb{Z}/m\mathbb{Z}$ has $\phi(m)$ invertible elements and $\mathbb{Z}/n\mathbb{Z}$ has $\phi(n)$, the number of such pairs is $\phi(m)\phi(n)$. □

In view of this corollary, it seems that the ring isomorphism version of the Chinese remainder theorem is a distinct improvement, since it unifies the classical theorem with the previously separate theorem that $\phi(mn) = \phi(m)\phi(n)$. Incidentally, the latter theorem is also enriched in this version – the equality of $\phi(mn)$ and $\phi(m)\phi(n)$ is actually a consequence of a *group isomorphism*, between the group of invertible elements of $\mathbb{Z}/mn\mathbb{Z}$ and the group of invertible elements of $(\mathbb{Z}/m\mathbb{Z}) \times (\mathbb{Z}/n\mathbb{Z})$. This isomorphism is simply what remains of the ring isomorphism after the noninvertible elements and the $+$ operation are forgotten.

Exercises

6.7.1 Check the ring properties (Section 2.1) of the direct product.

6.7.2 Show that solving the simultaneous congruences

$$x \equiv a \pmod{m}, \quad x \equiv b \pmod{n}$$

is equivalent to solving a linear equation in integers, which can be done by using the Euclidean algorithm to find $\gcd(m, n)$.

6.8 Homomorphisms and Quotient Rings

Mappings of rings that preserve $+$ and $\times$ are called *homomorphisms*. They need not by any means be one-to-one. An important type of example is the *evaluation map* that sends each polynomial $p(x)$ in a polynomial ring $F[x]$ to its value $p(\alpha)$ at $x = \alpha$. Evaluation obviously preserves $+$ and $\times$, but it is one-to-one only if α is transcendental over F, that is, if $p(\alpha) \neq 0$ for each $p(x) \in F[x]$. If $h(x) \in F[x]$ is a polynomial satisfied by α, then all multiples of $h(x)$, that is, all members of $h(x)F[x]$, are sent to 0 by the evaluation map.

We have met this situation previously (Section 4.5) when $h(x)$ is irreducible, in which case $h(x)F[x]$ consists of *all* polynomials that evaluate to 0, and there is a one-to-one correspondence between the congruence classes $h(x)F[x] + p(x)$ and the values $p(\alpha)$. Generalising this example, we can reinterpret any homomorphism ψ of a ring R as an isomorphism on the set of "congruence classes" of R modulo the class of elements that ψ sends to 0.

The technical language used to describe the general situation is the following. If $\psi : R \to R'$ is a homomorphism of ring R onto ring R' then the set

$$\psi^{-1}(0) = \{r \in R : \psi(r) = 0\}$$

is called the *kernel* of ψ. It is often denoted by $\ker \psi$. For each $a \in R$ we let

$$\psi^{-1}(0) + a = \{r + a : r \in \psi^{-1}(0)\} = \{r + a : \psi(r) = 0\}$$

and call it the *congruence class* of a (mod $\ker \psi$). It is the equivalence class of a under the relation

$$s \equiv t \pmod{\ker \psi}$$

defined by $s - t \in \ker \psi$.

Lemma. *If $a_1 \equiv b_1 \pmod{\ker \psi}$ and $a_2 \equiv b_2 \pmod{\ker \psi}$ then $a_1 + a_2 \equiv b_1 + b_2 \pmod{\ker \psi}$ and $a_1 a_2 \equiv b_1 b_2 \pmod{\ker \psi}$.*

Proof. This follows by the argument used in Section 2.6 to prove the corresponding properties of congruence mod n in $\mathbb{Z}$, except that instead of using the fact that the multiples of n are closed under sums and integer multiples we use the fact that $\ker \psi$ is closed under sums and multiples by members of R. Namely,

$$\begin{aligned}
r_1 \in \ker \psi,\ r_2 \in \ker \psi \quad &\Rightarrow \quad \psi(r_1) = 0,\ \psi(r_2) = 0 \\
&\Rightarrow \quad \psi(r_1 + r_2) = \psi(r_1) + \psi(r_2) = 0 + 0 = 0 \\
&\Rightarrow \quad r_1 + r_2 \in \ker \psi. \\
r \in \ker \psi \quad &\Rightarrow \quad \psi(r) = 0 \\
&\Rightarrow \quad \psi(kr) = \psi(k)\psi(r) = \psi(k) \times 0 = 0 \\
&\Rightarrow \quad kr \in \ker \psi \quad \text{for any } k \in R.
\end{aligned}$$

□

It follows from the lemma that $+$ and $\times$ are well defined on congruence classes by

$$(\psi^{-1}(0)+a)+(\psi^{-1}(0)+b)=\psi^{-1}(0)+(a+b)$$
$$(\psi^{-1}(0)+a)(\psi^{-1}(0)+b)=\psi^{-1}(0)+ab,$$

and that the congruence classes form a ring under these operations, since they inherit the ring properties from R (the same way $+$ and $\times$ are defined on congruence classes mod n and inherit the ring properties from $\mathbb{Z}$). The ring of congruence classes is denoted by $R/\ker\psi$ and we have:

Isomorphism Theorem for Rings. *If ψ is a homomorphism of ring R onto ring R' then the ring $R/\ker\psi$ is isomorphic to R'.*

Proof. The main step is to see that if $\psi(a) = a'$ then the congruence class $\psi^{-1}(0)+a$ of a is simply

$$\psi^{-1}(a')=\{s\in R:\psi(s)=a'\}.$$

On the one hand,

$$\begin{aligned} s\in\psi^{-1}(0)+a &\Rightarrow s=r+a \quad \text{where} \quad \psi(r)=0\\ &\Rightarrow \psi(s)=\psi(r)+\psi(a)=a'. \end{aligned}$$

On the other hand,

$$\begin{aligned} \psi(s)=a' &\Rightarrow \psi(s)=\psi(a)\\ &\Rightarrow 0=\psi(s)-\psi(a)=\psi(s-a)\\ &\Rightarrow s-a\in\psi^{-1}(0)\\ &\Rightarrow s\in\psi^{-1}(0)+a. \end{aligned}$$

Thus we have a one-to-one correspondence between classes $\psi^{-1}(a')\in R/\ker\psi$ and elements $a'\in R'$. This correspondence preserves $+$ and $\times$ because if $\psi(a)=a'$, $\psi(b)=b'$ then the class $\psi^{-1}(a'+b')$ is

$$\psi^{-1}(\psi(a)+\psi(b))=\psi^{-1}(\psi(a+b))=\psi^{-1}(0)+(a+b),$$

and similarly $\psi^{-1}(a'b')=\psi^{-1}(0)+ab$. □

Corollary. *A homomorphism $\psi: R\to R'$ is an isomorphism $\Leftrightarrow \ker\psi=\{0\}$.*

Proof. If $\ker\psi$ includes elements other than 0 then ψ is not one-to-one. Conversely, if $\ker\psi=\{0\}$ then the congruence class of each $a\in R$ consists of a alone, hence ψ is one-to-one. □

Remarks

(1) The only properties of $\ker\psi$ needed to show that sum and product of congruence classes mod $\ker\psi$ are meaningful are:

$$a,b\in\ker\psi\Rightarrow a+b\in\ker\psi,$$
$$a\in\ker\psi, k\in R\Rightarrow ka\in\ker\psi.$$

Thus if I is *any* subset of R closed under sums and under products with members of R we shall be able to define a ring R/I of congruence classes mod I. This is a useful point of view, though actually no more general, because any such subset is in fact the kernel of a homomorphism – the so-called *canonical* homomorphism $\psi : R \to R/I$ defined by $\psi(r) = I + r = \{i + r : i \in I\}$.

(2) The notation $\mathbb{Z}/n\mathbb{Z}$ we used in Section 2.7 for the ring of congruence classes mod n can now be seen as an instance of the notation R/I. Here we have $R = \mathbb{Z}$, $I = n\mathbb{Z} = \{nr : r \in \mathbb{Z}\}$, which is obviously closed under sums and under products with elements of $\mathbb{Z}$.

(3) The isomorphism theorem really is a theorem about rings because any homomorphism ψ of a field F onto a field F' is already an isomorphism. Just consider any nonzero $a \in F$. Then a^{-1} exists and the theorem in Section 6.1 shows

$$1 = \psi(1) = \psi(aa^{-1}) = \psi(a)\psi(a^{-1}),$$

hence $\psi(a) \neq 0$. Thus $\ker \psi = \{0\}$, and therefore ψ is an isomorphism by the corollary.

This fact helps to explain the overwhelming importance of isomorphisms in the theory of fields. While the structure of a ring can often be elucidated by a homomorphism onto a simpler ring – recall how we learned about the structure of $\mathbb{Z}$ in Chapter 2 by mapping it onto $\mathbb{Z}/n\mathbb{Z}$ – the only option for fields is to consider isomorphisms. For many fields, the automorphism group holds the key. As Galois discovered, groups are more tractable than fields because they too possess "simplifying" homomorphisms.

Exercises

6.8.1 Verify that if $I \subseteq R$ is closed under sums, and under products with elements of R, then the map $\psi(r) = I + r$ is a homomorphism with kernel I.

6.8.2 Find a natural homomorphism $\mathbb{Z} \to (\mathbb{Z}/m\mathbb{Z}) \times (\mathbb{Z}/n\mathbb{Z})$, and hence give another proof of the Chinese remainder theorem.

6.9 Discussion

The concepts of isomorphism and homomorphism emerged only gradually in algebra, being observed first for groups around 1830, for fields around 1870 and for rings around 1920. In his memoir on the solvability of equations, Galois [1831] implicitly analysed groups by means of homomorphisms. We shall say more about this in Chapters 7, 8 and 9. His groups are in fact automorphism groups of fields, though he did not see them that way, having constructed them directly from permutations of the roots of equations. The two examples he mentions specifically are the general equation of degree n, where the permutations are *all* permutations of the n roots, and the cyclotomic equation $\Phi_p(x) = 0$ for p prime, where he finds the $p-1$ automorphisms $\sigma_k(\zeta_p) = \zeta_p^k$ for $k = 1, \ldots, p-1$ (tacitly assuming Gauss's theorem that $\Phi_p(x)$ is irreducible).

The term "conjugation" was introduced by Cauchy [1821], Chapter VII, §1, for the isomorphism $a + bi \mapsto a - bi$ of $\mathbb{C}$.

The first to use the term "isomorphism" was Jordan, in his *Traité des Substitutions* [1870], the first textbook on group theory. As the title suggests, groups were then viewed concretely as groups of one-to-one functions (substitutions) under composition. The motivating examples, as for Galois, were permutations of the roots of equations. Jordan was strongly influenced by Galois and his book is a thorough development of the group theory only sketched or hinted at in Galois' work. Jordan used the word "isomorphism" for both isomorphisms and homomorphisms, but distinguished between the two by calling them "isomorphismes holoédriques" and "isomorphismes mériédriques" respectively.

In the 1870s Dedekind also began writing up his reflections on Galois theory (which he had lectured on as early as 1857/8) in his supplements to Dirichlet's *Vorlesungen über Zahlentheorie* ([1871] – [1893]). He realised that the group concept alone did not give a good explanation of Galois theory – it had to be complemented by the concept of field, which he was the first to identify and name. He chose the name "Körper," which also means "body" in German, because it "denotes a system with a certain completeness, fullness and self-containment" (Dedekind [1893], §160). He reinterpreted Galois' groups as automorphism groups of fields by observing that a permutation of the roots $\alpha_1, \ldots, \alpha_n$ of an equation extends to an automorphism of the field $\mathbb{Q}(\alpha_1, \ldots, \alpha_n)$ they generate. He even used the word "permutation" for any field isomorphism (Dedekind [1893], §161).

As mentioned in Section 2.9, the concept of ring emerged much later than this, in the 1920s, when it was finally recognised as a useful generalisation of integers, polynomials and algebraic integers. The concept is nevertheless implicit in Dedekind, who discovered the fundamental concept of *ideal* while investigating the integers of cyclotomic fields. The reason for the term "ideal" is a long story, but the concept turns out to be the same as the kernel of a ring homomorphism. The credit for recognising this, and the general importance of rings and homomorphisms, seems to be due to Emmy Noether.

Emmy Noether is one of the most influential figures in 20th century mathematics, but her influence is hard to trace precisely, due to her scarcity of publications and generosity in giving credit to others. She used to say: "Es steht alles schon bei Dedekind" ("Everything is already in Dedekind") – but it was her genius to make explicit the ideas which were only implicit in her predecessors (rather like Dedekind himself did). Her ideas became widely known mainly through van der Waerden's *Moderne Algebra* [1931], the first of the "groups, rings and fields" algebra books. Van der Waerden [1975] credited most of the ideas in the book to lectures of Emmy Noether in Göttingen 1924/25 and 1927/28, and lectures of Emil Artin in Hamburg 1926. Emmy Noether's important paper [1929], which contains the isomorphism theorem for rings, was also based on notes taken by van der Waerden at her course in 1927/28. A more general ring isomorphism theorem is given in Noether [1927], and she says it is implicit in the work of Sono [1917] (a hard-to-find Kyoto University publication I have not seen).

The Chinese remainder theorem is an excellent illustration of the power of the ring concept, but also of its slow emergence. Figure. 6.9.1 shows the first known statement of a Chinese remainder problem, in Sun Tzu's *Mathematical Classic*, around 300 AD. A full translation may be found in Libbrecht [1973], p.269, but the data is easily picked out from the two columns on the right (the text is read in columns, from right to left).

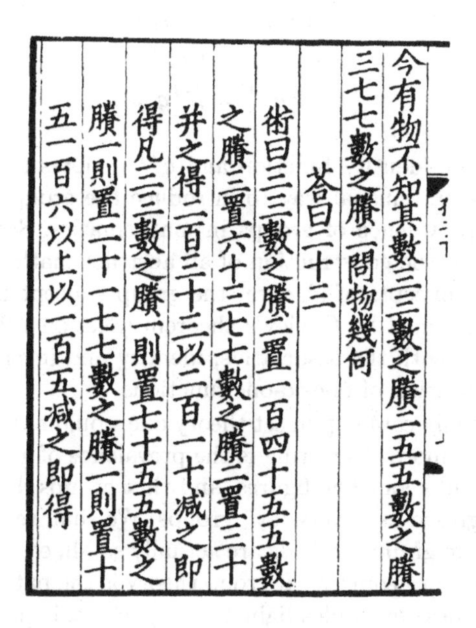
今有物不知其數三三數之賸二五五數之賸
三七七數之賸二問物幾何
荅曰二十三
術曰三三數之賸二置一百四十五五數
之賸三置六十三七七數之賸二置三十
并之得二百三十三以二百一十減之即
得凡三三數之賸一則置七十五五數之
賸一則置二十一七七數之賸一則置十
五一百六以上以一百五減之即得

Fig. 6.9.1. The Chinese remainder problem

"Counting by threes (indicated by writing the character 三 for 3 twice), the remainder is 2 (二), counting by fives (五 五), the remainder is 3 (三), counting by sevens (七七), the remainder is (二)." The third column from the right gives the answer 23 (二十三).

The Chinese understood how to reduce such problems to linear equations, and how to solve them using the Euclidean algorithm. The connection with the Euler phi function seems to have gone unnoticed until Gauss [1801], article 38, used a one-to-one correspondence between the integers $x < mn$ and relatively prime to mn and pairs (y, z) with y relatively prime to m and z relatively prime to n to prove

$$\phi(mn) = \phi(m)\phi(n)$$

when m is relatively prime to n. The correspondence was probably not observed to be a ring isomorphism until the 1920s, but I have been unable to find exactly when this happened.

7 Groups

7.1 Why Groups?

The concept of group came up briefly in Section 6.4, when we observed the following three properties of one-to-one functions from a set to itself:

$$g_1(g_2g_3) = (g_1g_2)g_3 \qquad (\textit{associativity})$$
$$g\mathbf{1} = \mathbf{1}g \qquad (\textit{identity})$$
$$gg^{-1} = g^{-1}g = \mathbf{1} \qquad (\textit{inverse})$$

where $\mathbf{1}$ denotes the identity function and g^{-1} denotes the inverse of function g. This is perhaps an unusual way to introduce the group concept, but it does help explain the prevalence of groups in mathematics. Sets and functions are the raw material for the construction of all abstract mathematical objects and, among the major concepts of algebra, the group concept is closest to pure set theory. This is confirmed by Cayley's theorem (Section 7.2), which shows that any binary relation with the associative, identity and inverse properties can be modelled by composition of one-to-one functions.

Still, if groups only came up in set theory one would not expect to meet them very often. What is more interesting is the prevalence of groups at the everyday level of numbers and geometric figures, and that groups defined at the set level can often be recognised as "everyday" groups. We shall see in this chapter that many automorphism groups of fields are actually small, easily understood groups from number theory or geometry. If group theory is the path to enlightenment in field theory, then these examples light the way. For this reason, we shall devote most of the chapter to examples, proving only such theorems as are absolutely fundamental.

The definition of group that encompasses all examples is the following. A *group* is a set G together with a function $G \times G \to G$, denoted by juxtaposition, an *identity element* $\mathbf{1} \in G$, and for each $g \in G$ an *inverse* $g^{-1} \in G$, with the properties

$$g_1(g_2g_3) = (g_1g_2)g_3 \text{ for all } g_1, g_2, g_3 \in G,$$
$$g\mathbf{1} = \mathbf{1}g = g \text{ for all } g \in G,$$
$$gg^{-1} = g^{-1}g = \mathbf{1} \text{ for all } g \in G.$$

The reason for using this notation is of course to conform with the example of function composition, or with the example of number multiplication, which is where the notation originated. This notation, called *multiplicative notation*, is not always the most convenient. Its main rival is *additive notation*, in which the binary operation is written as $+$, the identity element as $\mathbf{0}$, and the inverse of g as $-g$. In additive notation the group properties are therefore

$$g_1 + (g_2 + g_3) = (g_1 + g_2) + g_3$$
$$g + \mathbf{0} = \mathbf{0} + g = g$$
$$g + (-g) = (-g) + g = \mathbf{0}.$$

Since number multiplication and addition have the properties just enumerated, it follows that any set of numbers closed under $\times$ and $\div$ is a group under multiplication, and any set of numbers closed under $+$ and $-$ is a group under addition.

For example, $\mathbb{Q}-\{0\}, \mathbb{R}-\{0\}$ and $\mathbb{C}-\{0\}$ are groups under $\times$; $\mathbb{Z}, \mathbb{Q}, \mathbb{R}$ and $\mathbb{C}$ are groups under $+$. These are certainly "everyday" groups, but they are atypical in some respects.

In the first place they are infinite, and all the automorphism groups we shall meet are finite. In the second place they are *abelian*, that is, $g_1g_2 = g_2g_1$ (in the multiplicative notation) or $g_1+g_2 = g_2+g_1$ (in the additive notation) for all g_1, g_2 in the group. Abelian groups are quite important – in fact the additive notation is often reserved for them – but they are not typical of group theory, which is largely about coping with *non*abelian operations. We shall therefore have to look a little beyond ordinary arithmetic for instructive "everyday" groups.

The nearest source of finite abelian groups is in arithmetic mod n, as we shall see in Section 7.3. The nearest source of finite nonabelian groups is in geometry, as we shall see in Section 7.4.

Exercises

7.1.1 Give more concise definitions of ring and field (Section 2.2), in terms of the concept of group.

7.1.2 Is $\mathbb{Z}$ a group under subtraction?

7.2 Cayley's Theorem

We now wish to back up the claim from Section 7.1 that any group can be modelled by a group of one-to-one functions under composition. The functions are constructed explicitly from the group elements to produce a group isomorphic to the original, in the appropriate sense of "isomorphism" for groups. An *isomorphism* ψ of group G onto group G' is a one-to-one function that preserves the group operation, that is,

$$\psi(g_1g_2) = \psi(g_1)\psi(g_2) \quad \text{for all } g_1, g_2 \in G.$$

(It follows that ψ also preserves the identity element and inverses; see Exercise 7.2.1 which follows.)

The functions corresponding to group elements are very natural and may be illustrated by the group $\mathbb{Z}$ under addition. Corresponding to each $m \in \mathbb{Z}$ is the one-to-one function $m+$ that adds m to each $l \in \mathbb{Z}$, namely,

$$m + (l) = m + l.$$

The functions $m+$ are different for different m, and the correspondence preserves sums because the composition of $n+$ with $m+$ is the function

$$n + m + (l) = (n + m) + l = (n + m) + (l)$$

that adds $n+m$ to each $l \in \mathbb{Z}$. The proof of Cayley's theorem is a straightforward generalisation of this idea. The only point where the general case requires more thought is in showing that the function $g\times$ corresponding to $g \in G$ (we will use multiplicative notation now) is one-to-one and onto G. This follows from the group properties of G.

Cayley's Theorem. *Any group G is isomorphic to a group G' of one-to-one functions from G onto itself.*

Proof. Let each $g \in G$ correspond to the "left multiplication by g" function $g\times$ defined by

$$g \times (h) = gh \quad \text{for all } h \in G.$$

Then we have a correspondence $\psi(g) = g\times$ between G and the set G' of left multiplication functions. The correspondence is one-to-one because if $g_1 \neq g_2$ the functions $g_1\times$ and $g_2\times$ are different, for example $g_1 \times (1) \neq g_2 \times (1)$. And the correspondence preserves products because

$$g_1 \times (g_2 \times (h)) = g_1 g_2 h = g_1 g_2 \times (h),$$

that is,

$$\psi(g_1)\psi(g_2) = \psi(g_1 g_2).$$

Thus the group G is isomorphic to the group G' of functions under composition – if G' is indeed a group. This actually follows on general grounds (see the Exercise) but it can also be seen directly because:

(i) each $g\times$ in G' is a one-to-one function from G onto G.

The function $g\times$ is one-to-one because

$$g \times (h_1) = g \times (h_2) \Rightarrow gh_1 = gh_2 \Rightarrow h_1 = h_2$$

by cancellation of g. It is onto G because each $g_1 \in G$ is $g \times (g^{-1}g_1)$.

(ii) G' is closed under inverses.

In fact the inverse of $g\times$ is $g^{-1}\times$, because

$$g^{-1} \times g \times (h) = h = \mathbf{1} \times h.$$

(iii) G' is closed under products.

As we have already seen, $g_1 \times g_2 \times (h) = g_1 g_2 \times h$. □

Exercise

7.2.1 Suppose G is a group and ψ maps G onto a set G' so that

$$\psi(g_1 g_2) = \psi(g_1)\psi(g_2),$$

where juxtaposition of $\psi(g_1)$ and $\psi(g_2)$ denotes a certain binary operation on G'. Verify that G' is a group under this operation, with $\psi(\mathbf{1})$ the identity element and $\psi(g^{-1})$ the inverse of $\psi(g)$.

7.3 Abelian Groups

A group G is *abelian* if

$$g_1 g_2 = g_2 g_1 \quad \text{for all } g_1, g_2 \in G.$$

The examples of $\mathbb{Z}$, $\mathbb{Q}$, $\mathbb{R}$ and $\mathbb{C}$ under addition, mentioned in Section 7.1, are all infinite abelian groups. Important examples of *finite* abelian groups are contained in the finite rings and fields $\mathbb{Z}/n\mathbb{Z}$ and $\mathbb{Z}/p\mathbb{Z}$ (Section 2.7), and in the roots of unity.

The elements of $\mathbb{Z}/n\mathbb{Z}$ form an abelian group under addition, called the *cyclic group* C_n. Imagine the congruence classes $n\mathbb{Z}, n\mathbb{Z}+1, \ldots, n\mathbb{Z}+n-1$ arranged in a circle, then addition of $n\mathbb{Z}+m$ corresponds to rotation through m places. An isomorphic form of C_n is the set $\{1, \zeta_n, \ldots, \zeta_n^{n-1}\}$ under multiplication, where $\zeta_n = \cos\frac{2\pi}{n} + i\sin\frac{2\pi}{n}$ is an nth root of 1 (Section 3.7). The special case $n = 2$ gives C_2 as the group $\{1, -1\}$ under multiplication. In general, the group elements lie on the unit circle in the plane $\mathbb{C}$, and multiplication by ζ_n^m really rotates the elements through m places. Figure 7.3.1 shows this form of C_n for $n = 5$.

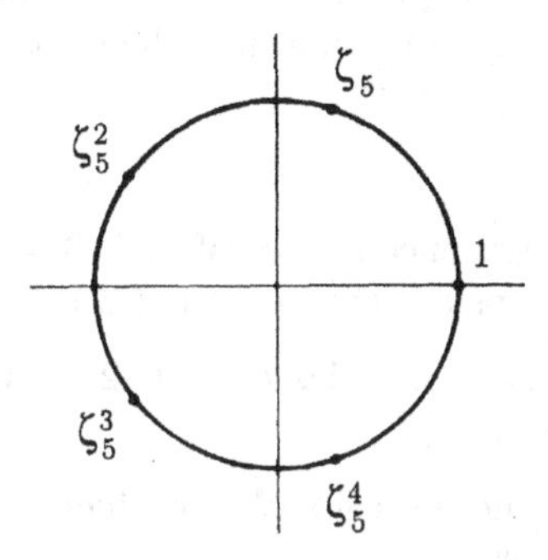

Fig. 7.3.1. Elements of C_5

The invertible elements of $\mathbb{Z}/n\mathbb{Z}$ form an abelian group under multiplication, which we denote by $(\mathbb{Z}/n\mathbb{Z})^\times$. We noticed in Section 6.6* that there is a one-to-one correspondence between the invertible elements k of $\mathbb{Z}/n\mathbb{Z}$ and the automorphisms $\sigma_k(\zeta_n) = \zeta_n^k$ of $\mathbb{Q}(\zeta_n)$. Now that we have group structure in mind, it is also easy to notice the following:

Theorem. *The automorphism group of $\mathbb{Q}(\zeta_n)$ is isomorphic to $(\mathbb{Z}/n\mathbb{Z})^\times$.*

Proof. It suffices to show that the product of the automorphisms $\sigma_i(\zeta_n) = \zeta_n^i$ and $\sigma_j(\zeta_n) = \zeta_n^j$ corresponding to $i, j \in \mathbb{Z}/n\mathbb{Z}$ is the automorphism $\sigma_{ij}(\zeta_n) = \zeta_n^{ij}$ corresponding to ij. Indeed it is, because

$$\sigma_i(\sigma_j(\zeta_n)) = \sigma_i(\zeta_n^j) = (\sigma_i(\zeta_n))^j = (\zeta_n^i)^j = \zeta_n^{ij}. \qquad \square$$

In the special case where n is a prime p, $(\mathbb{Z}/p\mathbb{Z})^\times$ happens to be cyclic, though the proof is far from obvious (Section 8.7). The difficulty lies in finding a number whose powers mod p include all of $1, 2, \ldots, p-1$. (See also Exercises 7.3.1 and 7.3.2.)

Given two abelian groups A, B, a third is obtained as their *direct product*

$$A \times B = \{(a, b) : a \in A, b \in B\}$$

under the obvious group operation on pairs:

$$(a_1, b_1)(a_2, b_2) = (a_1 a_2, b_1 b_2).$$

If $\mathbf{1}_A$, $\mathbf{1}_B$ are the respective identity elements of A, B then $(\mathbf{1}_A, \mathbf{1}_B)$ is the identity of $A \times B$. Also, the inverse of (a, b) is (a^{-1}, b^{-1}).

The simplest nontrivial direct product is $C_2 \times C_2$, and it is not isomorphic to a cyclic group (Exercise 7.3.3). $C_2 \times C_2$ also arises as the automorphism group of $\mathbb{Q}(\sqrt{2}, \sqrt{3})$. We know from Section 6.3 that the automorphisms σ of $\mathbb{Q}(\sqrt{2}, \sqrt{3})$ correspond to the four sets of values $\sigma(\sqrt{2}) = \pm\sqrt{2}$, $\sigma(\sqrt{3}) = \pm\sqrt{3}$. In other words, σ does two things: it multiplies $\sqrt{2}$ by ± 1, and it multiplies $\sqrt{3}$ by ± 1. We can therefore represent σ by the pair (a, b) of multipliers. It is easy to see that the product of automorphisms then corresponds to the product of pairs defined above, hence the group of automorphisms is isomorphic to the group of pairs $(\pm 1, \pm 1)$, that is, to $C_2 \times C_2$.

Exercises

7.3.1 Show that the congruence classes of 1, 2, 3, 4 mod 5 form a cyclic group under multiplication. (Consider the powers of 2.)

7.3.2 Show that the congruence classes of 1, 2, 3, 4, 5, 6 mod 7 form a cyclic group under multiplication.

7.3.3 Show that there is no $g \in C_2 \times C_2$ for which $1, g, g^2, g^3$ are distinct, hence that $C_2 \times C_2$ is not cyclic.

7.3.4 Show that $C_2 \times C_3$ is isomorphic to C_6.

7.3.5 Under what conditions is $C_m \times C_n$ cyclic?

7.3.6 Give examples of n for which $(\mathbb{Z}/n\mathbb{Z})^\times$ is not cyclic.

7.4 Dihedral Groups

The simplest nonabelian groups are the *dihedral groups* D_n for $n \geq 3$. D_n is the symmetry group of the regular n-gon, that is, the group of motions of the n-gon that map it onto the same region in space. These include the n rotations through multiples of $2\pi/n$ about the center of the n-gon; also rotation through π about an axis of symmetry, which exchanges the front and back of the n-gon. When this exchange is followed by a rotation, n new positions are attainable; hence D_n has $2n$ elements. Since both front and back come into view under these motions, the n-gon is sometimes called a *dihedron* (two faces), and this is why the group is called dihedral.

The n-gon is considered to exist only for $n \geq 3$, hence D_3 is the smallest dihedral group. If r denotes the clockwise rotation through $2\pi/3$, and f (for "flip") the rotation through π about the vertical axis of symmetry, then the six elements of D_3 can be represented by the corresponding positions of an equilateral triangle ABC whose initial position is labelled **1** (Figure 7.4.1. The shading represents the back of the triangle.) Note that rf must be interpreted "f, then r." We write products of group elements in right to left order because we wish to view them as functions. Order is now important because $fr \neq rf$. By applying f to the picture for r one sees in fact that $fr = r^2 f$.

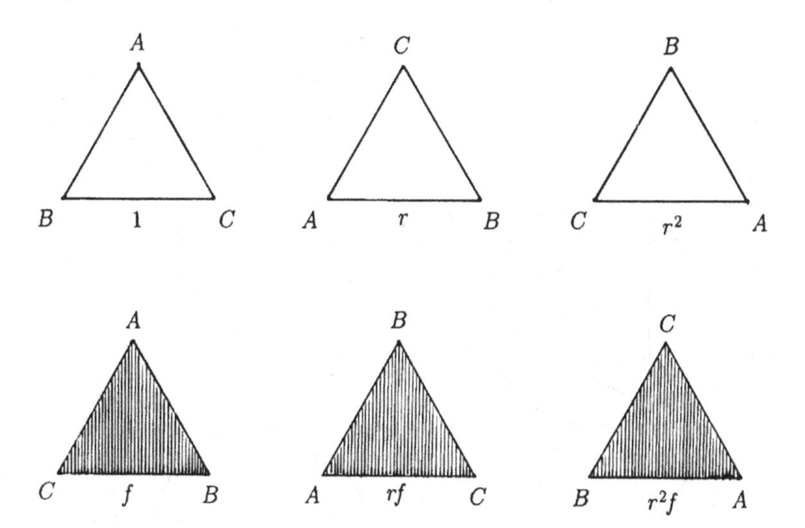

Fig. 7.4.1. Symmetries of a triangle

A similar argument shows that $fr = r^{n-1}f \neq rf$ in each D_n, hence each D_n is a nonabelian group.

Some groups D_n arise quite naturally as automorphism groups of fields.

Example. D_3 is the automorphism group of $\mathbb{Q}(\sqrt[3]{2}, \zeta_3)$.

Since $\mathbb{Q}(\sqrt[3]{2}, \zeta_3)$ consists of rational functions of $\sqrt[3]{2}$ and ζ_3, an automorphism σ of $\mathbb{Q}(\sqrt[3]{2}, \zeta_3)$ is determined by the values of $\sigma(\sqrt[3]{2})$ and $\sigma(\zeta_3)$.

Also $(\sigma(\sqrt[3]{2}))^3 = \sigma(2) = 2$, hence $\sigma(\sqrt[3]{2})$ is $\sqrt[3]{2}$, $\zeta_3\sqrt[3]{2}$ or $\zeta_3^2\sqrt[3]{2}$ (the three cube roots of 2). And $(\sigma(\zeta_3))^3 = \sigma(1) = 1$, hence $\sigma(\zeta_3)$ is ζ_3 or ζ_3^2 (the two cube roots of 1 other than 1). This gives six possible σ which in fact are composed from the following σ_1 and σ_2:

$$\sigma_1(\sqrt[3]{2}) = \zeta_3\sqrt[3]{2},\ \sigma_1(\zeta_3) = \zeta_3,$$
$$\sigma_2(\sqrt[3]{2}) = \sqrt[3]{2},\ \sigma_2(\zeta_3) = \zeta_3^2.$$

It was found in Section 6.2 that σ_1 and σ_2 *are* automorphisms, so it follows that there are six automorphisms of $\mathbb{Q}(\sqrt[3]{2},\zeta_3)$. We now have to set up a correspondence between these six automorphisms and the symmetries of an equilateral triangle. The vertices of the triangle are staring us in the face: $A = \sqrt[3]{2}$, $B = \zeta_3\sqrt[3]{2}$ and $C = \zeta_3^2\sqrt[3]{2}$. Computing $\sigma_1(A), \sigma_1(B), \sigma_1(C)$, we find that σ_1 *rotates* A, B, C; moving A to B, B to C and C to A. Similarly, we find that σ_2 *flips* the triangle around the real axis by exchanging B and C (Figure 7.4.2). Thus the six automorphisms of $\mathbb{Q}(\sqrt[3]{2},\zeta_3)$ correspond to the six symmetries of the triangle ABC. Since products are preserved by definition of the correspondence, we have an isomorphism as claimed. □

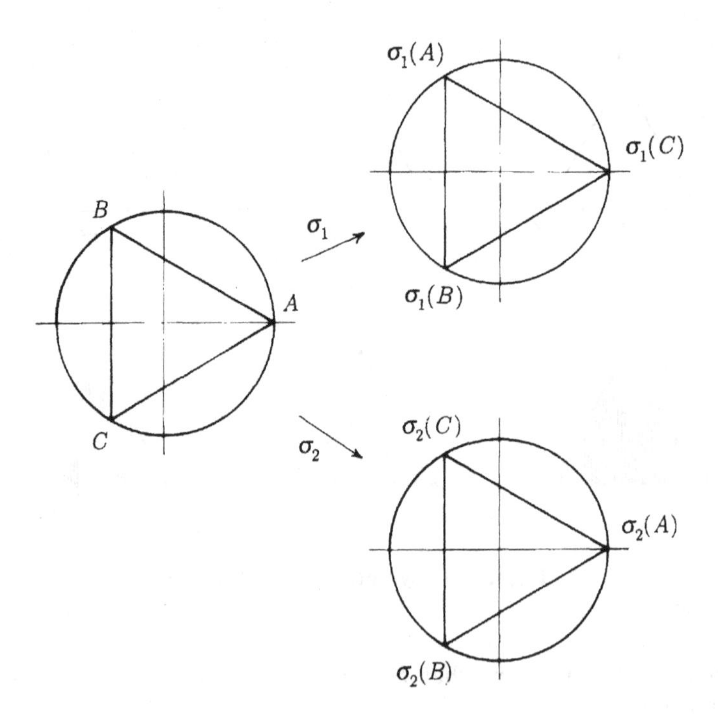

Fig. 7.4.2. Automorphisms of $\mathbb{Q}(\sqrt[3]{2},\zeta_3)$

Exercise

7.4.1 Show that the automorphisms of $\mathbb{Q}(\sqrt[4]{2},i)$ (previously encountered in Exercise 6.2.1) form a group isomorphic to the symmetry group of a square.

7.5* Permutation Groups

Permutation is just another word for a one-to-one function from a set to itself, most often used when the set is finite. The group of all $n!$ permutations of an n-element set is called the *symmetric group* and is denoted by S_n. These permutations came up in Section 6.5* as the automorphisms of $\mathbb{Q}(x_1,\ldots,x_n)$ that fix the elementary symmetric functions $a_0,\ldots,a_{n-1}$ of $x_1,\ldots,x_n$, with the permuted objects being the n indeterminates $x_1,\ldots,x_n$. Thus the theorem in Section 6.5* can be restated as the following theorem about groups.

Theorem. *The automorphisms of $\mathbb{Q}(x_1,\ldots,x_n)$ fixing $a_0,\ldots,a_{n-1}$ form a group isomorphic to S_n.* □

S_n was first studied because of its property of fixing the symmetric functions of $x_1,\ldots,x_n$. An important group contained in S_n is the *alternating group* A_n, most concisely defined as the group of permutations that fix the function

$$\Delta(x_1,\ldots,x_n)=\prod_{i<j}(x_i-x_j).$$

This function is not symmetric, and hence A_n is not all of S_n.

An element of S_n not in A_n is the permutation σ that exchanges x_1 and x_2 and leaves the remaining x_i fixed. This permutation replaces the factor (x_1-x_2) of $\Delta(x_1,\ldots,x_n)$ by $(x_2-x_1)=-(x_1-x_2)$ and exchanges each other (x_1-x_i) with (x_2-x_i), thus replacing $\Delta(x_1,\ldots,x_n)$ by $-\Delta(x_1,\ldots,x_n)$. The sign change can be traced to the single factor, namely (x_1-x_2), whose subscripts are reversed in order by σ. In general, a permutation σ reverses the sign of a factor (x_i-x_j) if the subscripts of $\sigma(x_i)$ and $\sigma(x_j)$ are opposite in order to i, j. Hence $\Delta(x_1,\ldots,x_n)$ is fixed if σ reverses the order of subscripts in an even number of factors, and otherwise $\Delta(x_1,\ldots,x_n)$ is replaced by $-\Delta(x_1,\ldots,x_n)$. In the first case we say σ is an *even* permutation, and in the second case we say it is *odd.* Thus A_n is the group of even permutations of $x_1,\ldots,x_n$.

Forgetting about the letter x, we can say that a permutation σ of $\{1,2,\ldots,n\}$ is even if it produces an even number of order reversals, that is, if there are an even number of pairs i,j such that $i<j \Leftrightarrow \sigma(i)>\sigma(j)$. This gives a quick diagrammatic way to recognise evenness of permutations. Just write the sequence $1,2,\ldots,n$ twice, in parallel rows, and draw a line from each i in the top row to $\sigma(i)$ in the bottom row. Then σ is even if and only if the lines have an even number of crossings.

For example, the "3-cycle" σ of $\{1,2,3\}$ defined by $\sigma(1)=2$, $\sigma(2)=3$, $\sigma(3)=1$ is even because its diagram (Figure 7.5.1) has two crossings. The reason that this works in general is that each crossing corresponds to an order reversal.

Exercises

7.5.1 Show that each $\sigma\in S_n$ is a product $\tau_1\tau_2\ldots\tau_r$ where each τ_i is a *transposition* – a permutation that exchanges two elements and fixes the rest. (Hint: $\tau_1,\ldots,\tau_r$ can be read off the diagram of σ.)

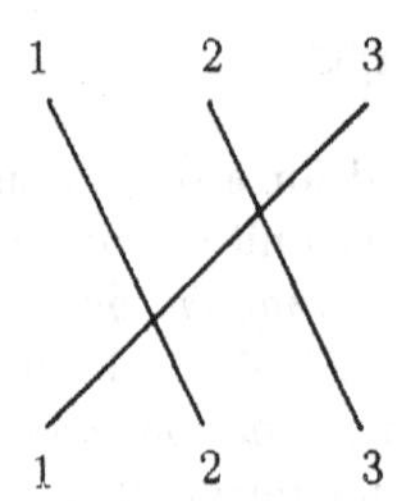

Fig. 7.5.1. Diagram of a 3-cycle

7.5.2 If $n \geq 3$, show that a product of two transpositions in S_n is also a product of 3-cycles. Deduce that the elements of A_n are precisely the products of 3-cycles.

7.5.3 Show that each transposition (i, j) is a product of transpositions of the form $(k, k+1)$.

7.5.4 Show that each transposition $(k, k+1)$ is a product of (1,2) and powers of the n-cycle $(1, 2, \ldots, n)$.

7.6* Permutation Groups in Geometry

It follows from the proof of Cayley's theorem (Section 7.2) that any group G of n elements is isomorphic to a group of permutations of n things, and hence contained in S_n. This is an occasionally useful insight but not a very deep one, since the n things are simply the members of G. It also doesn't tell us that S_n is anything but a general purpose container for finite groups. We might learn a lot more by looking for permutations which represent G more economically, in other words, seek the smallest S_m containing a copy of G. This idea pays off handsomely for the symmetry groups G of the regular solids, revealing that the first few S_n and A_n actually belong to geometry.

The first interesting case is S_3, which turns out to be the same as the dihedral group D_3. This can be seen from the six positions of the triangle in Figure 7.4.1, which correspond to the six permutations of the vertex set $\{A, B, C\}$. Since the product of motions corresponds to the product of permutations, this correspondence is an isomorphism between D_3 and S_3.

This discovery prompts us to investigate how rotations of the tetrahedron (Figure 7.6.1) permute its vertex set $\{A, B, C, D\}$. A rotation through $2\pi/3$ about the vertical line through A leaves A fixed and cycles B, C and D, hence it is an even permutation. By combining this rotation with the analogous rotation that fixes B and cycles C, D and A, one eventually obtains all 12 even permutations of $\{A, B, C, D\}$. But *only* 12 positions of the tetrahedron are obtainable by rotation, since a position is determined when one knows which of the four faces is on the bottom (say) and which of its three vertices is in front (say). Thus we have a one-to-one correspondence between positions and even permutations of $\{A, B, C, D\}$, hence *the group of rotations of the tetrahedron is isomorphic to* A_4.

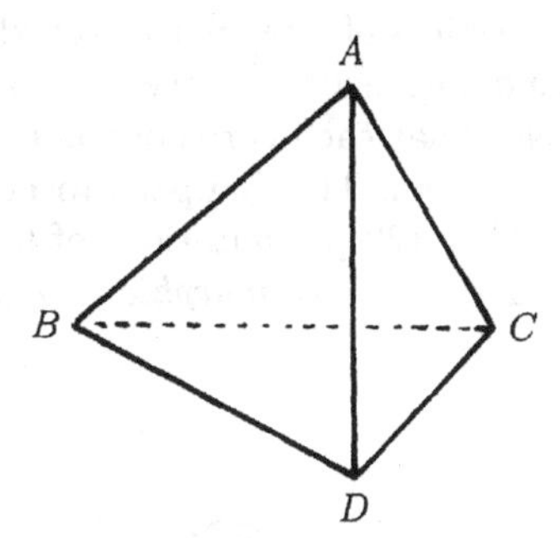

Fig. 7.6.1. Tetrahedron

The cube (Figure 7.6.2) has eight vertices, but it is not necessary to represent its rotations by permutations of eight things. Rotations yield only $6 \times 4 = 24$ positions, determined by which of the six faces is on the bottom and which of its four vertices is (say) at the left front. Since $24 = 4!$ this is actually the number of permutations of *four* things, and four things materialise as the four pairs of opposite vertices, marked by A, B, C, D in Figure 7.6.2. Again one eventually finds, by rotating the cube enough, that each position of the cube corresponds to a different permutation of $\{A, B, C, D\}$. Thus *the group of rotations of the cube is isomorphic to* S_4. (To test this, think of a random permutation of A, B, C, D and see whether you can find your permutation on the boundary of one of the faces in Figure 7.6.2.)

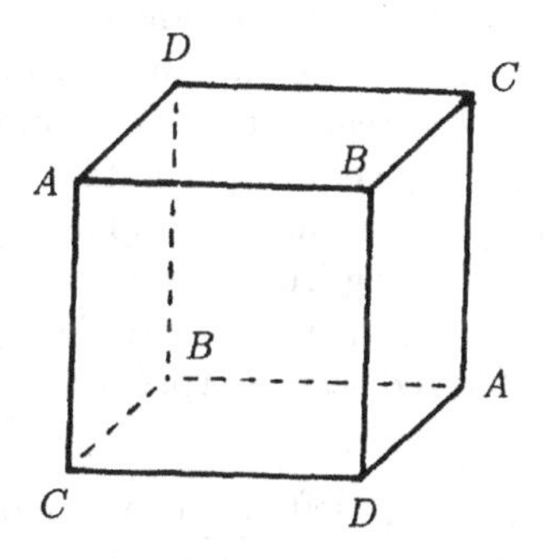

Fig. 7.6.2. Cube

Finally, consider the dodecahedron (Figure 7.6.3). It has 20 vertices, which may be partitioned into five *tetrahedral sets*, like the four points marked A in Figure 7.6.3 (which are the vertices of a regular tetrahedron). There are $12 \times 5 = 60$ possible positions of the dodecahedron, since there are twelve pentagonal faces. It turns out, miraculously, that each corresponds to a *different*, *even* permutation of the five tetrahedral sets. This happens to be all the even permutations of five things – half the $5! = 120$ permutations of five things – hence *the group of rotations of the dodecahedron is isomorphic to* A_5.

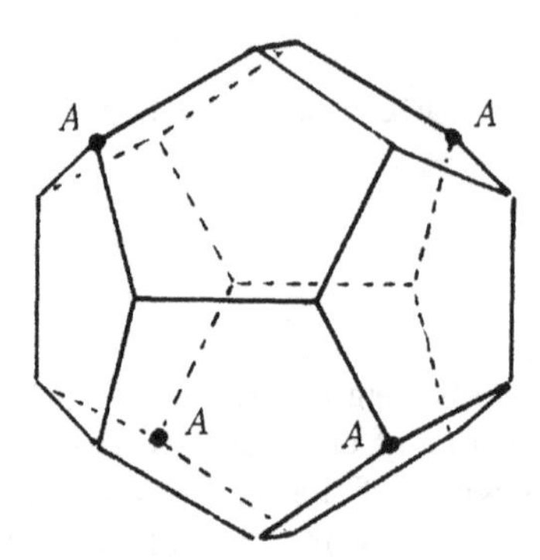

Fig. 7.6.3. Dodecahedron

Remarks. It can be left to the reader to check that exactly half the permutations of three or five things are even, since this is a routine computation. However, there is a much shorter general explanation why A_n is half the size of S_n, which will be given in the next section. It is also possible to explain the appearance of all the permutations (in A_4, S_4 and A_5 respectively) more briefly than actually finding them all (on the tetrahedron, cube and dodecahedron respectively). However, in this case the explanation has no useful generalisation, so we shall leave the interested reader to look it up in Lamotke [1986] p.18, or Coxeter [1961], p.273. The discovery of permutation groups in geometry seems to have occurred as a spinoff from more complicated investigations of polynomial functions in geometry, arising from the representation of the sphere as the completion of $\mathbb{C}$ by the point ∞ (see Klein [1876]). Several more years passed before Klein [1884] presented a direct interpretation of A_4, S_4 and A_5 as groups of rotations.

The function $\Delta(x_1, x_2, x_3, x_4)$ left fixed by A_4 is very helpful in analysing nested square roots of the form $\sqrt{r + s\sqrt{t}}$, where $r, s, t \in \mathbb{Q}$. In particular, it gives a proof that the side $\frac{1}{5}\sqrt{10(5 - \sqrt{5})}$ of the icosahedron inscribed in the unit sphere is not of the form $\sqrt{a} \pm \sqrt{b}$ for $a, b \in \mathbb{Q}$ (see Exercises 7.6.3 – 7.6.6).

Exercises

7.6.1 The rotation group of the octahedron is isomorphic to S_4 and the rotation group of the icosahedron is isomorphic to A_5. Explain why.

7.6.2 Suppose the faces of an octahedron are coloured black and white alternately, that is, so that each edge separates a black face from a white face. Show that the group of rotations which preserve the face colours is isomorphic to A_4.

7.6.3 If $\alpha = \sqrt{r + s\sqrt{t}}$ with $r, s, t \in \mathbb{Q}$, let $\alpha' = \sqrt{r - s\sqrt{t}}$ and show that

$$\begin{aligned}\Delta(\alpha, \alpha', -\alpha, -\alpha') &= 4\alpha\alpha'(\alpha - \alpha')^2(\alpha + \alpha')^2 \\ &= 2^4 s^2 t\sqrt{r^2 - s^2 t}\end{aligned}$$

and hence that

$$\Delta(\alpha, \alpha', -\alpha, -\alpha') \in \mathbb{Q} \ \Leftrightarrow \ \sqrt{r^2 - s^2 t} \in \mathbb{Q}.$$

7.6.4 Show, on the other hand, that if $a, b \in \mathbb{Q}$ then

$$\Delta(\sqrt{a} + \sqrt{b}, \sqrt{a} - \sqrt{b}, -\sqrt{a} - \sqrt{b}, -\sqrt{a} + \sqrt{b}) \in \mathbb{Q}.$$

7.6.5 Deduce from Exercises 7.6.3, 7.6.4 and 4.2.6 that a root $\sqrt{r + s\sqrt{t}}$ of an irreducible quartic $f(x) \in \mathbb{Q}[x]$ is of the form $\sqrt{a} \pm \sqrt{b}$ (where $a, b, r, s, t \in \mathbb{Q}$) only if $\sqrt{r^2 - s^2 t} \in \mathbb{Q}$.

7.6.6 Conclude, using Exercise 4.7.7, that the side of the icosahedron inscribed in the unit sphere is not of the form $\sqrt{a} \pm \sqrt{b}$ for $a, b \in \mathbb{Q}$.

7.7 Subgroups and Cosets

A group H contained in a group G is called a *subgroup of* G. A subgroup H decomposes G into sets of the form

$$gH = \{gh : h \in H\} \quad \text{where } g \in G$$

called *left cosets* of H. Since G may not be abelian we also have to consider *right cosets*

$$Hg = \{hg : h \in H\},$$

though usually one can stick to one type or the other.

A simple but important example is the subgroup $n\mathbb{Z} = \{nk : k \in \mathbb{Z}\}$ of $\mathbb{Z}$ (with addition as the group operation). The cosets of $n\mathbb{Z}$ in $\mathbb{Z}$ are nothing but the congruence classes mod n : $n\mathbb{Z}, n\mathbb{Z} + 1, \ldots, n\mathbb{Z} + n - 1$ (Section 2.7). They are written in additive notation, and as right cosets, though in this case the right coset $n\mathbb{Z} + g$ equals the left coset $g + n\mathbb{Z}$ since $+$ is commutative.

Cosets help us to dissect the structure of G because of the following theorem.

Theorem. *For any $g_1, g_2 \in G$ the cosets g_1H, g_2H of a subgroup H are of equal size and are either identical or disjoint.*

Proof. The coset $g_1H = \{g_1h : h \in H\}$ is mapped onto $g_2H = \{g_2h : h \in H\}$ by the function $g_2g_1^{-1}\times$ (compare with Section 7.2). This function must be one-to-one since it is inverted by the function $g_1g_2^{-1}\times : g_2H \to g_1H$. Thus there is a one-to-one correspondence between g_1H and g_2H and hence they are the same size.

If g_1H, g_2H have a common element g then $g = g_1h_1$ for some $h_1 \in H$ and also $g = g_2h_2$ for some $h_2 \in H$. Thus $g_1h_1 = g_2h_2$ and therefore $g_1 = g_2h_2h_1^{-1}$ (multiplying both sides on the right by h_1^{-1}). It follows that

$$g_1H = g_2h_2h_1^{-1}H = g_2(h_2h_1^{-1}H) = g_2H$$

since $h_2h_1^{-1} \in H$ and therefore $h_2h_1^{-1}H = H$ (recall from Cayley's theorem that left multiplication by a group element maps the group onto itself). □

A famous corollary to this theorem is the result known as *Lagrange's theorem:*

Lagrange's Theorem. *If H is a subgroup of a finite group G then the size of H divides the size of G.*

Proof. Denote the size of G by $|G|$ and the size of H by $|H|$. By the theorem, G is partitioned into disjoint cosets of size $|H|$, hence $|H|$ divides $|G|$. □

We can use the partition into equal sized cosets to prove that A_n is half the size of S_n, as claimed in Section 7.6. The relevant property of A_n is that the product of odd permutations σ_1, σ_2 is even, since if σ_1, σ_2 both change the sign of $\Delta(x_1, \ldots, x_n)$ then $\sigma_1\sigma_2$ fixes it. It follows that any odd permutation σ can be written as the product of a fixed odd permutation τ and the even permutation $\tau^{-1}\sigma$, hence $\sigma \in \tau A_n$. Thus A_n has only two left cosets in S_n, A_n itself and τA_n, and since these have equal size, A_n is half the size of S_n. The size $|G|$ of a group G is also called the *order* of G.

The number $|G|/|H|$ of cosets of a subgroup H of a finite group G is called the *index* of H in G and denoted by $[G : H]$. The index has a multiplicative property like the dimension $(E : F)$ of a field E over a subfield F (Section 5.4).

Index Product Theorem. *If $G \supseteq H \supseteq K$ are groups with G finite, then the index $[G : K] = [G : H][H : K]$.*

Proof. $[G : K] = |G|/|K| = (|G|/|H|)(|H|/|K|) = [G : H][H : K]$. □

Exercises

7.7.1 If G is any finite group and $g \in G$, show that the powers $\mathbf{1}, g, g^2, \ldots$ form a subgroup of G. (The size of this subgroup, which is the least n such that $g^n = \mathbf{1}$, is called the *order* of g.)

7.7.2 Deduce from Exercise 7.7.1 that the least n such that $g^n = \mathbf{1}$, is a divisor of $|G|$. Also that $g^{|G|} = \mathbf{1}$.

7.7.3 Deduce from Exercises 7.7.1 and 7.7.2 that if $|G|$ is a prime p then G is isomorphic to C_p.

7.7.4 By considering the group G of nonzero elements of $\mathbb{Z}/p\mathbb{Z}$ under $\times$, deduce from Exercise 7.7.2 that $a \not\equiv 0 \pmod p \Rightarrow a^{p-1} \equiv 1 \pmod p$ (another proof of Fermat's little theorem, see Section 2.8*).

7.7.5 Is there a similar explanation of Euler's theorem? (Section 2.8*)

7.8 Normal Subgroups

If H is a subgroup of G we know from Section 7.7 that G is partitioned into cosets $g_1H, g_2H, \ldots$. Reflecting on the special case where $G = \mathbb{Z}$, $H = n\mathbb{Z}$ and the cosets are the congruence classes mod n, we might attempt to define a *product of cosets* by

$$g_iH \cdot g_jH = g_ig_jH$$

(analogous to the sum of congruence classes). If meaningful, this product inherits the group properties from G and makes the set of cosets into a group, called the *quotient* G/H of G by H. In particular, the identity element of G/H is the coset $1 \cdot H = H$. However, the product of cosets is *not* always well defined.

If it is, consider the *coset map* $\psi(g) = gH$ from G to G/H. The elements sent to H by ψ are precisely the $h \in H$. On the other hand, ghg^{-1} goes to

$$gH \cdot hH \cdot g^{-1}H = gH \cdot H \cdot g^{-1}H = gH \cdot g^{-1}H = gg^{-1}H = H,$$

so $ghg^{-1} \in H$ for all $g \in G, h \in H$. Making the obvious abbreviation gHg^{-1} for $\{ghg^{-1} : h \in H\}$, we find

$$\begin{aligned} ghg^{-1} \in H \text{ for all } g \in G, h \in H &\Rightarrow gHg^{-1} \subseteq H \text{ for all } g \in G \\ &\Rightarrow gH \subseteq Hg, \end{aligned}$$

multiplying both sides on the right by g. Repeating the argument with g^{-1} in place of g yields $Hg \subseteq gH$, hence in fact H has the property that $gH = Hg$ for all $g \in G$. Such a subgroup is called *normal.*

Conversely, if H is normal and $g_1H = g_1'H$, $g_2H = g_2'H$ then

$$\begin{aligned} g_1g_2H &= g_1g_2'H && \text{since } g_2H = g_2'H \\ &= g_1Hg_2' && \text{since } g_2'H = Hg_2' \\ &= g_1'Hg_2' && \text{since } g_1H = g_1'H \\ &= g_1'g_2'H && \text{since } Hg_2' = g_2'H. \end{aligned}$$

Thus the product $g_iH \cdot g_jH = g_ig_jH$ is well defined. We have proved the theorem:

Theorem. *H is a normal subgroup of G if and only if the cosets gH for $g \in G$ form a group G/H under the product*

$$g_iH \cdot g_jH = g_ig_jH. \qquad \square$$

The problem of non-normal subgroups only arises when G is a nonabelian group, which is also when we have the most trouble dissecting the structure of G.

In the groups that arise in Chapter 8 we shall be able to deal with this problem by finding subgroups H "large enough" to make G/H abelian.

Examples from our present repertoire are the groups $D_3 = S_3$ and A_4.

As we saw in Section 7.4, D_3 is not abelian since $rf \neq fr$. However, the subgroup $C_3 = \{\mathbf{1}, r, r^2\}$ contains half of D_3 and therefore its only other left coset is $fC_3 = \{f, fr, fr^2\} = D_3 - C_3$. Since the right coset $C_3 f$ also equals $D_3 - C_3$, we necessarily have $fC_3 = C_3 f$. In fact *any* coset of C_3 is either C_3 or $D_3 - C_3$, so $gC_3 = C_3 g$ for any $g \in D_3$. Thus C_3 is a normal subgroup of D_3 and D_3/C_3 has two elements, which means that D_3/C_3 is isomorphic to the abelian group C_2 (for example by Exercise 7.7.3).

This argument shows, more generally, that any subgroup H which is half the size of a finite group G is a normal subgroup of G, since its only cosets are H and $G - H$. In particular, each A_n is a normal subgroup of S_n, and S_n/A_n is isomorphic to C_2.

Finding proper normal subgroups of A_n is more of a challenge, and indeed we succeed only with A_4. It has a normal subgroup isomorphic to $C_2 \times C_2$, consisting of the identity and the three permutations that simultaneously transpose two pairs (the latter correspond to half turns of the tetrahedron fixing the midpoints of opposite edges). Since A_4 has twelve elements and $C_2 \times C_2$ has four, it follows that $A_4/(C_2 \times C_2)$ has three elements, so it is necessarily isomorphic to C_3.

Exercises

7.8.1 Show that the subgroup $\{\mathbf{1}, f\}$ of D_n is not normal for all $n \geq 3$, and give examples of cosets whose product is not well defined.

7.8.2 Check that the copy of $C_2 \times C_2$ in A_4, described above, is normal.

7.8.3 Show that A_4 has no subgroup with six elements (which shows, incidentally, that the converse of Lagrange's theorem is false).

7.8.4 If H is a subgroup with only two cosets in an infinite group G, is H normal?

7.9 Homomorphisms

A map ψ from a group G onto a group G' is called a *homomorphism* if it preserves products, that is, if

$$\psi(g_1 g_2) = \psi(g_1)\psi(g_2) \quad \text{for all } g_1, g_2 \in G.$$

Immediate consequences of this definition are:
(i) ψ preserves $\mathbf{1}$ because

$$\psi(g) = \psi(g \cdot \mathbf{1}) = \psi(g)\psi(\mathbf{1}) \Rightarrow \mathbf{1} = \psi(\mathbf{1}),$$

multiplying both sides by $\psi(g)^{-1}$ on the left,
(ii) ψ preserves inverses because

$$\mathbf{1} = \psi(\mathbf{1}) = \psi(gg^{-1}) = \psi(g)\psi(g^{-1}) \Rightarrow \psi(g^{-1}) = \psi(g)^{-1}.$$

The prototype of a homomorphism is the *coset map* $\psi : G \to G/H$ defined by $\psi(g) = gH$ when H is a normal subgroup of G. The coset map preserves products because

$$\psi(g_1g_2) = g_1g_2H = g_1Hg_2H = \psi(g_1)\psi(g_2)$$

by definition of the product of cosets. In fact, any homomorphism ψ of G onto G' can be viewed as a coset map if we take H to be the *kernel* of ψ:

$$\ker \psi = \{g \in G : \psi(g) = \mathbf{1}\}.$$

This emerges from the proof of the following theorem, which is the group analogue of the theorem about rings in Section 6.8.

Isomorphism Theorem for Groups. *If ψ is a homomorphism of group G onto group G' then $\ker \psi$ is a normal subgroup of G and $G/\ker\psi$ is isomorphic to G'.*

Proof. Since ψ is onto G' the elements $g' \in G'$ are in one-to-one correspondence with the following subsets of G:

$$\psi^{-1}(g') = \{g \in G : \psi(g) = g'\}.$$

In particular, $\mathbf{1} \in G'$ corresponds to

$$\psi^{-1}(\mathbf{1}) = \{g \in G : \psi(g) = \mathbf{1}\} = \ker\psi.$$

And if $\psi(g) = g'$ then

$$\psi^{-1}(g') = (\ker\psi)g = g(\ker\psi)$$

because

$$\begin{aligned} g^* \in (\ker\psi)g &\Leftrightarrow g^*g^{-1} \in \ker\psi \\ &\Leftrightarrow \psi(g^*g^{-1}) = \mathbf{1} \\ &\Leftrightarrow \psi(g^*)\psi(g^{-1}) = \mathbf{1} \quad \text{since } \psi \text{ is a homomorphism} \\ &\Leftrightarrow \psi(g^*)\psi(g)^{-1} = \mathbf{1} \quad \text{since } \psi \text{ preserves inverses} \\ &\Leftrightarrow \psi(g^*) = \psi(g) = g' \\ &\Leftrightarrow g^* \in \psi^{-1}(g'), \end{aligned}$$

and similarly,

$$g^* \in g(\ker\psi) \Leftrightarrow g^* \in \psi^{-1}(g').$$

Thus $\ker\psi$ is a normal subgroup of G and its cosets are the sets $\psi^{-1}(g')$ for $g' \in G'$. Finally, the correspondence $g' \mapsto \psi^{-1}(g')$ preserves products because

$$\psi^{-1}(g_1'g_2') = g_1g_2(\ker\psi) = g_1(\ker\psi)\cdot g_2(\ker\psi) = \psi^{-1}(g_1')\psi^{-1}(g_2'),$$

hence it is an isomorphism of G' onto $G/\ker\psi$. □

Some of the normal subgroups mentioned in Section 7.8 appear naturally as kernels of homomorphisms. At the same time, of course, their cosets appear as the inverse images of nonidentity elements.

Example 1. The permutation sign homomorphism.

A_n is the kernel of the *sign* homomorphism $S_n \to C_2 = \{1, -1\}$ that sends a permutation σ to -1 if σ changes the sign of $\Delta(x_1, \ldots, x_n)$ and to 1 if it does not (compare with the definition of A_n in Section 7.5). Thus each S_n (for $n \geq 2$) has a homomorphism onto C_2. The set of odd permutations, the inverse image of -1, is the other coset of A_n. □

The purpose of homomorphisms, as we mentioned when ring homomorphisms came up in Section 6.8, is to "simplify" a structure by mapping it onto smaller structures. By mapping a group G onto a smaller group G' by a homomorphism we simplify G by "factoring out" the normal subgroup $\ker \phi$. Of course, finding ϕ is theoretically equivalent to finding $\ker \phi$, but in an actual situation the homomorphism of G may be more obvious than the normal subgroup. One such situation is the following.

Example 2. The restriction homomorphism.

G is a group of automorphisms of a field F, and $F' \subseteq F$ is a field mapped onto itself by each $\sigma \in G$. Then each $\sigma \in G$ has a *restriction to* F' – the map $\sigma' = \sigma|_{F'}$ of F' that agrees with σ there – which is an automorphism of F', and the restrictions σ' form a group G' of automorphisms of F'. The *restriction map* $|_F : G \to G'$ has the homomorphism property $\sigma_1\sigma_2|_{F'} = \sigma_1|_{F'}\sigma_2|_{F'}$ because $\sigma_1|_{F'}\sigma_2|_{F'}(f')$ exists for each $f' \in F'$, since each σ maps F' onto itself, and obviously

$$\sigma_1|_{F'}\sigma_2|_{F'}(f') = \sigma_1\sigma_2(f') = \sigma_1\sigma_2|_{F'}(f').$$

Thus G' is a "simplification" of G obtained by factoring out the kernel of the restriction homomorphism. This homomorphism is one of the keys to the relationship between field structure and group structure, which will be investigated in Chapters 8 and 9. For the moment we shall just illustrate how it works with the group G of automorphisms of $F = \mathbb{Q}(\sqrt[3]{2}, \zeta_3)$.

As we have seen in Section 7.4, G is generated by the automorphisms σ_1 and σ_2, where $\sigma_1(\sqrt[3]{2}) = \zeta_3\sqrt[3]{2}$, $\sigma_1(\zeta_3) = \zeta_3$ and $\sigma_2(\sqrt[3]{2}) = \sqrt[3]{2}$, $\sigma_2(\zeta_2) = \zeta_2^2$ Both of these map the subfield $F' = \mathbb{Q}(\zeta_3)$ onto itself. Thus we have a restriction homomorphism $|_{F'}$ sending σ_1, σ_2 respectively to the automorphisms $\sigma_1'(\zeta_3) = \zeta_3$ (the identity) and $\sigma_2'(\zeta_3) = \zeta_3^2$ of F'. The latter automorphisms form a group G' isomorphic to C_2, since $(\sigma_2')^2$ is the identity. This is confirmed by factoring G by the kernel of $|_{F'}$, which is indeed the normal subgroup $\{1, \sigma_1, \sigma_1^2\}$ – the automorphisms of F that fix ζ_3 (and hence all elements of F') – with quotient isomorphic to C_2. □

The use of homomorphisms is often more efficient than the use of normal subgroups and their cosets. Compare the two approaches in the following case (which will come up again in Chapter 8).

Example 3. Abelian quotient groups.

We shall prove: *if G/H is abelian then $x, y \in G \Rightarrow x^{-1}y^{-1}xy \in H$.*

(i) Proof using cosets.

Since G/H is the group of cosets gH for $g \in G$, and since G/H is abelian, we have

$$\begin{aligned} x^{-1}y^{-1}xyH &= x^{-1}Hy^{-1}HxHyH \quad \text{by definition of the product of cosets} \\ &= x^{-1}HxHy^{-1}HyH \quad \text{since the cosets commute} \\ &= x^{-1}xy^{-1}yH \quad \text{by definition of the product of cosets} \\ &= 1H \\ &= H. \end{aligned}$$

Hence $x^{-1}y^{-1}xy \in H$.

(ii) Proof using a homomorphism.
Since G/H is abelian, there is a homomorphism $\psi : G \to$ abelian group, with kernel H.

$$\begin{aligned} \psi(x^{-1}y^{-1}xy) &= \psi(x)^{-1}\psi(y)^{-1}\psi(x)\psi(y) \quad \text{since } \psi \text{ is a homomorphism} \\ &= \psi(x)^{-1}\psi(x)\psi(y)^{-1}\psi(y) \quad \text{since the } \psi \text{ values commute} \\ &= 1. \end{aligned}$$

Hence $x^{-1}y^{-1}xy \in \ker \psi = H$. □

Exercises

7.9.1 What relationships between automorphism groups are revealed by restricting the automorphisms of $\mathbb{Q}(\sqrt{2}, \sqrt{3})$ to $\mathbb{Q}(\sqrt{2})$?

7.9.2 Consider the effect of restricting the automorphisms of $\mathbb{Q}(\zeta_n)$ to $\mathbb{Q}(\cos \frac{2\pi}{n})$ (compare with Exercise 6.6.3). What is the kernel of the restriction map?

7.10 Discussion

The group concept has historical origins in number theory and geometry as well as in the theory of equations. For example, we have seen how Euler [1761] used inverses mod p to prove Fermat's little theorem (Section 2.8*). In geometry, Euler [1776] proved that the product of two rotations in $\mathbb{R}^3$ is again a rotation, thus paving the way for studying the symmetry of regular solids via their rotation groups. However, the importance of inverses and closure under products was not recognised until late in the 19th century, when the general group concept was finally identified. Between 1770 and 1870 group theory was almost entirely the theory of finite permutation groups arising in the theory of equations.

This theory was started by Lagrange in his [1771], a lengthy analysis of the nth degree equation and the effect of permutations on rational functions of its roots. Lagrange realised, without explicitly formulating the group concept, that understanding the general equation of degree n depended on understanding S_n and its subgroups. He saw that S_n is the group fixing the symmetric functions of the roots $x_1, \ldots, x_n$, and that other functions of $x_1, \ldots, x_n$, arising in the process of solving the equation, are fixed by subgroups. It was left to Galois to discover the precise subgroup structure that enables solution by radicals, as we shall see in Chapter 8. However, Lagrange managed to explain the known solutions of cubic and quartic equations (Sections 1.6 and 1.7) in terms of the structure of S_3 and S_4. In particular, he discovered the exceptional nature of A_4 and its subgroup isomorphic to $C_2 \times C_2$. He viewed the latter as the group fixing the function $x_1x_2 + x_3x_4$.

In the course of these investigations Lagrange discovered that the size of any subgroup of S_n divides the size $n!$ of S_n. What we now call Lagrange's theorem is actually due to Jordan [1870], p.25. Jordan generously attributed the theorem to Lagrange after observing that Lagrange's proof (partitioning into cosets) applies equally well to any finite group.

The function $\Delta(x_1, \ldots, x_n) = \prod_{i<j}(x_i - x_j)$ fixed by A_n is a square root of a function called the *discriminant.* Since the permutations in S_n that change the sign of $\Delta(x_1, \ldots, x_n)$ leave its square fixed, the discriminant is fixed by all permutations in S_n. It follows, by the fundamental theorem of symmetric functions (Section 6.5*), that the discriminant is a polynomial in the coefficients $a_0, \ldots, a_{n-1}$ of

$$x^n + a_{n-1}x^{n-1} + \cdots + a_1x + a_0 = (x - x_1)\cdots(x - x_n).$$

For example, the discriminant of the quadratic

$$\begin{aligned} x^2 + a_1x + a_0 &= (x - x_1)(x - x_2) \\ &= x^2 - (x_1 + x_2)x + x_1x_2 \end{aligned}$$

is

$$(x_1 - x_2)^2 = (x_1 + x_2)^2 - 4x_1x_2 = a_1^2 - 4a_0,$$

as we have already seen in Section 1.6. The discriminant is so called because it discriminates between equations with repeated roots (discriminant $= 0$) and those with distinct roots (discriminant $\neq 0$), as is obvious from the definition of $\Delta(x_1, \ldots, x_n)$. The result just mentioned means that we can decide whether a given equation has repeated roots by computing a polynomial in its coefficients.

Functions such as $\Delta(x_1, \ldots, x_n)$, which undergo only a sign change (at most) under permutations of $x_1, \ldots, x_n$ were called *alternating functions* by Cauchy [1815]. In this paper Cauchy began to isolate the group properties of permutations, defining the product of permutations and the identity permutation. Among the products he singled out the powers of a permutation, and showed that each permutation has a power equal to the identity.

The latter property was taken as an axiom by Cayley [1854], in the first attempt to define the abstract group concept. Cayley assumed a finite set of elements, closed under an associative operation he wrote as product, with an identity element he wrote as 1, and such that each element θ satisfies $\theta^n = 1$ for some n. This of course implies that θ has the inverse θ^{n-1}, but the concept of inverse apparently did not yet suggest itself as a useful primitive notion.

The concept of inverse was still lurking below the surface when Kronecker [1870] wrote down axioms for finite abelian groups. Kronecker's axioms were associativity, commutativity, identity and – instead of inverse – cancellation. His statement of cancellation is that

$$\theta' \neq \theta'' \Rightarrow \theta\theta' \neq \theta\theta'' \text{ for any } \theta,$$

which of course is equivalent to

$$\theta\theta' = \theta\theta'' \Rightarrow \theta' = \theta'' \text{ for any } \theta.$$

This implies that any θ has an inverse, as follows. The sequence $\theta, \theta^2, \theta^3, \ldots$ must eventually contain a θ^s equal to an earlier θ^r, by finiteness (this was Cauchy's argument), in which case $\theta^{s-r} = 1$ by cancellation. Thus cancellation implies Cayley's axiom, and hence an inverse of θ.

As mentioned in Section 7.3, Kronecker's paper contains a proof that every finite abelian group is isomorphic to a direct product of cyclic groups. This was the first theorem about groups directly inspired by number theory. Kronecker needed the theorem to analyse finite abelian groups arising in number theory, in his case the so-called ideal class groups, but the groups $(\mathbb{Z}/n\mathbb{Z})^\times$ are an equally good example. The latter are not obviously direct products of cyclic groups, but in fact they have cyclic factors related to the prime factors of n (see for example Ireland and Rosen [1982], p.44).

In the 1870s geometry also began to influence group theory. The discovery that the polyhedral groups are A_4, S_4 and A_5 was made by Klein [1876], but more important was his *Erlanger Programm* [1872], which emphasised the unifying role of groups in geometry. Groups turn up everywhere in geometry – groups of motions, projections and homeomorphisms for example – and they are mostly *infinite* groups. This brought the concept of inverse out into the open, because the cancellation property is not sufficient to guarantee the existence of inverses in an infinite associative system (for example $\mathbb{N}$ under addition). The existence of inverses was finally stated explicitly by Klein's student Dyck in Dyck [1883], a paper which grew out of the study of symmetries of infinite tessellations.

As mentioned in Section 6.9, the concepts of isomorphism and homomorphism were introduced by Jordan [1870]. In fact, immediately after introducing them, on p.56 of his book, Jordan essentially proves that the kernel of a homomorphism is a normal subgroup. He therefore comes very close to the isomorphism theorem for groups, but it seems that this theorem (like the corresponding one for rings) was not clearly formulated until the 1920s.

8 Galois Theory of Unsolvability

8.1 Galois Groups

In Chapters 6 and 7 we observed that the automorphisms of any field form a group, reflecting the "algebraic symmetry" of the field. In many cases we were able to recognise the group in question, by its abstract structure, as one of the standard groups from number theory or geometry. Some of the groups identified were the following (with sections where the identification was made shown at the right):

Field	Automorphism group	Section
$\mathbb{Q}(\sqrt{2})$	C_2	(6.4)
$\mathbb{Q}(\sqrt[3]{2})$	$\{1\}$	(6.2)
$\mathbb{Q}(\zeta_n)$	$(\mathbb{Z}/n\mathbb{Z})^\times$	(6.6*)
$\mathbb{Q}(\sqrt{2}, \sqrt{3})$	$C_2 \times C_2$	(7.3)
$\mathbb{Q}(\sqrt[3]{2}, \zeta_3)$	D_3	(7.4)

We have also generalised this idea to the automorphisms of a field E that fix a subfield F, thus obtaining a group which reflects the symmetry of E relative to F. In Sections 6.5* and 7.5* we saw that S_n is the group of automorphisms of $\mathbb{Q}(x_1, \ldots, x_n)$ which fix the elementary symmetric functions $a_0, \ldots, a_{n-1}$ of $x_1, \ldots, x_n$ and hence the subfield $\mathbb{Q}(a_0, \ldots, a_{n-1})$. In Section 7.5* we also defined A_n to be the subgroup of S_n fixing $\prod_{i<j}(x_i - x_j)$, which means that A_n is the group of automorphisms of $\mathbb{Q}(x_1, \ldots, x_n)$ fixing the subfield $\mathbb{Q}(a_0, \ldots, a_{n-1}, \prod_{i<j}(x_i - x_j))$.

In general, we say that an automorphism $\sigma : E \to E$ fixes a field $F \subseteq E$ if $\sigma(a) = a$ for all $a \in F$. Since $\sigma(a) = a$ implies $a = \sigma^{-1}(a)$, the inverse of an automorphism fixing F also fixes F. And if $\sigma_1(a) = a$ and $\sigma_2(a) = a$ then $\sigma_1\sigma_2(a) = a$, hence if σ_1 and σ_2 fix F so does $\sigma_1\sigma_2$. Thus the set of automorphisms σ of E that fix F is closed under inverses and products and therefore is a group. It is called the *Galois group of E over F* and we denote it by $\mathrm{Gal}(E : F)$. (Many authors denote it by $\mathrm{Gal}(E/F)$, making use of the mathematical pun "E/F" for "E over F." I believe it is already confusing enough to have the / sign used both for quotients of numbers and quotients of groups or rings.)

All the automorphism groups in the table above are Galois groups over $\mathbb{Q}$, since $\mathbb{Q}$ is fixed by any automorphism of a field $E \supseteq \mathbb{Q}$ (Corollary 6.1). Thus we can rewrite the results in the table, using the $\cong$ sign for isomorphism, as

$$\begin{aligned}
\mathrm{Gal}(\mathbb{Q}(\sqrt{2}) : \mathbb{Q}) &\cong C_2, \\
\mathrm{Gal}(\mathbb{Q}(\sqrt[3]{2}) : \mathbb{Q}) &\cong \{1\}, \\
\mathrm{Gal}(\mathbb{Q}(\zeta_n) : \mathbb{Q}) &\cong (\mathbb{Z}/n\mathbb{Z})^\times, \\
\mathrm{Gal}(\mathbb{Q}(\sqrt{2}, \sqrt{3}) : \mathbb{Q}) &\cong C_2 \times C_2, \\
\mathrm{Gal}(\mathbb{Q}(\sqrt[3]{2}, \zeta_3) : \mathbb{Q}) &\cong D_3,
\end{aligned}$$

and the results of Sections 6.5* and 7.5* are

$$\begin{aligned}
\mathrm{Gal}(\mathbb{Q}(x_1,\ldots,x_n):\mathbb{Q}(a_0,\ldots,a_{n-1})) &\cong S_n,\\
\mathrm{Gal}(\mathbb{Q}(x_1,\ldots,x_n):\mathbb{Q}(a_0,\ldots,a_{n-1},\prod_{i<j}(x_i-x_j))) &\cong A_n.
\end{aligned}$$

The latter results may be expressed by saying that $\mathbb{Q}(x_1,\ldots,x_n)$ has S_n symmetry over $\mathbb{Q}(a_0,\ldots,a_{n-1})$, but only A_n symmetry over its extension $\mathbb{Q}(a_0,\ldots,a_{n-1},\prod_{i<j}(x_i-x_j))$. The Galois group detects a "halving" of relative symmetry when the element $\prod_{i<j}(x_i-x_j)$ is adjoined.

Similar reductions in relative symmetry are detected by the Galois group when elements are adjoined to the bottom field $\mathbb{Q}$ in the earlier examples.

Example 1. $\mathrm{Gal}(\mathbb{Q}(\sqrt{2},\sqrt{3}):\mathbb{Q}) \cong C_2\times C_2$, $\mathrm{Gal}(\mathbb{Q}(\sqrt{2},\sqrt{3}):\mathbb{Q}(\sqrt{2})) \cong C_2$.

Of the four automorphisms σ of $\mathbb{Q}(\sqrt{2},\sqrt{3})$, found in Section 7.3, only the two with $\sigma(\sqrt{2})=\sqrt{2}$ fix $\mathbb{Q}(\sqrt{2})$. Hence $\mathrm{Gal}(\mathbb{Q}(\sqrt{2},\sqrt{3}):\mathbb{Q}(\sqrt{2}))$ has two members and it is necessarily isomorphic to C_2. □

Example 2. $\mathrm{Gal}(\mathbb{Q}(\sqrt[3]{2},\zeta_3):\mathbb{Q}) \cong D_3$, $\mathrm{Gal}(\mathbb{Q}(\sqrt[3]{2},\zeta_3):\mathbb{Q}(\zeta_3)) \cong C_3$.

Of the six automorphisms σ of $\mathbb{Q}(\sqrt[3]{2},\zeta_3)$, found in Section 7.4, only the three with $\sigma(\zeta_3)=\zeta_3$ fix $\mathbb{Q}(\zeta_3)$. Hence $\mathrm{Gal}(\mathbb{Q}(\sqrt[3]{2},\zeta_3):\mathbb{Q}(\zeta_3))$ has three members and it is necessarily isomorphic to C_3. □

These examples suggest that when E results from F by adjoining m^{th} roots, the group $\mathrm{Gal}(E:F)$ is in some sense "easily decomposed." The symmetry of E relative to F breaks down in small, easy stages as m^{th} roots are adjoined. We shall make this idea precise in Section 8.3, after discussing its bearing on the problem of solution by radicals in Section 8.2.

Proving that the general n^{th} degree equation is not solvable by radicals, when $n\geq 5$, is of course one of the main goals of this book. The ability of the Galois group to reflect adjunction of radicals is crucial to the proof. However, only the *concept* of Galois group is involved, not the theory of polynomials and irreducibility used to compute some of the specific Galois groups above. The little group theory we need can be developed from scratch using just the fundamental properties of normal subgroups and homomorphisms from Sections 7.8 and 7.9.

Exercises

8.1.1 Find the Galois groups of $\mathbb{Q}(\sqrt[4]{2},i)$ over $\mathbb{Q}$, $\mathbb{Q}(\sqrt{2})$ and $\mathbb{Q}(i)$.

8.1.2 Find $\mathrm{Gal}(\mathbb{Q}(\sqrt[3]{2},\zeta_3):\mathbb{Q}(\sqrt[3]{2}))$.

8.1.3 Interpret $\mathrm{Gal}(\mathbb{Q}(\sqrt[3]{2},\zeta_3):\mathbb{Q}(\zeta_3))$ as the group of rotations of a triangle.

8.2 Solution by Radicals

Recall from Section 1.8 that the general problem of solution by radicals was to express the roots of

$$x^n + a_{n-1}x^{n-1} + \cdots + a_1x + a_0 = 0 \qquad (*)$$

in terms of the coefficients $a_0, \ldots, a_{n-1}$ using a finite number of operations $+, -, \times, \div$ and $\sqrt{\;}, \sqrt[3]{\;}, \sqrt[4]{\;}, \ldots$. The elements obtained from $a_0, \ldots, a_{n-1}$ by the operations $+, -, \times, \div$ form the *coefficient field* $\mathbb{Q}(a_0, \ldots, a_{n-1})$. An element obtained by a finite number of operations $\sqrt{\;}, \sqrt[3]{\;}, \sqrt[4]{\;}, \ldots$ lies in an extension field of $\mathbb{Q}(a_0, \ldots, a_{n-1})$ obtained by a finite number of *radical adjunctions.* We say that adjunction of an element α to a field F is *radical* if there is a positive integer m such that $\alpha^m = f \in F$, in which case α may be denoted by the radical expression $\sqrt[m]{f}$. The result $F(\alpha_1)\ldots(\alpha_k) = F(\alpha_1, \ldots, \alpha_k)$ of the k radical adjunctions α_i to $F(\alpha_1, \ldots, \alpha_{i-1})$ is called a *radical extension* of F. Thus the problem of solution by radicals is to find a radical extension of the coefficient field $\mathbb{Q}(a_0, \ldots, a_{n-1})$ which includes the roots $x_1, \ldots, x_n$ of (*), and hence contains the *root field* $\mathbb{Q}(x_1, \ldots, x_n)$.

For example, the formula for the solution of the general quadratic equation (Section 1.5)

$$x_1, x_2 = \frac{-a_1 \pm \sqrt{a_1^2 - 4a_0}}{2}$$

shows that $\mathbb{Q}(x_1, x_2)$ is contained in the radical extension $\mathbb{Q}(a_0, a_1, \sqrt{a_1^2 - 4a_0})$. In fact, adjoining the radical $\sqrt{a_1^2 - 4a_0}$ to $\mathbb{Q}(a_0, a_1)$ gives *exactly* the field $\mathbb{Q}(x_1, x_2)$. On the other hand, solution of the general cubic equation (Section 1.6) gives a proper extension of the root field involving imaginary cube roots of unity, due to the presence of cube roots in the formula. For this reason, we should not expect a radical extension of $\mathbb{Q}(a_0, \ldots, a_{n-1})$ to equal $\mathbb{Q}(x_1, \ldots, x_n)$ – the question is whether it can contain it.

Further light is thrown on the problem by viewing $a_0, \ldots, a_{n-1}$ as functions of indeterminates $x_1, \ldots, x_n$, defined by the identity

$$(x - x_1) \cdots (x - x_n) = x^n + a_{n-1}x^{n-1} + \cdots + a_1x + a_0. \qquad (**)$$

As the reader may recall from Section 6.5*, $a_0, \ldots, a_{n-1}$ are polynomials in $x_1, \ldots, x_n$ called the *elementary symmetric functions*:

$$a_0 = (-1)^n x_1 \cdots x_n,$$

$$\vdots$$

$$a_{n-1} = -(x_1 + \cdots + x_n)$$

By viewing the general n^{th} degree polynomial in x as the left-hand side of (**), rather than the right-hand side, the problem becomes: extend the field $\mathbb{Q}(a_0, \ldots, a_{n-1})$ of functions $a_0 = (-1)^n x_1 \cdots x_n, \ldots, a_{n-1} = -(x_1 + \cdots + x_n)$

by radicals until it includes $x_1, \ldots, x_n$. When the solution of the quadratic is viewed in this way, it becomes obvious why the radical $\sqrt{a_1^2 - 4a_0}$ is important, and why $\mathbb{Q}(x_1, x_2) = \mathbb{Q}(a_0, a_1, \sqrt{a_1^2 - 4a_0})$. Namely,

$$\sqrt{a_1^2 - 4a_0} = \sqrt{(x_1 + x_2)^2 - 4x_1x_2} = \sqrt{(x_1 - x_2)^2} = \pm(x_1 - x_2).$$

Thus $\sqrt{a_1^2 - 4a_0} \in \mathbb{Q}(x_1, x_2)$. Conversely, both $x_1, x_2 \in \mathbb{Q}(a_0, a_1, \sqrt{a_1^2 - 4a_0})$, since the latter equals $\mathbb{Q}(x_1x_2, x_1 + x_2, x_1 - x_2)$.

Since $a_0, \ldots, a_{n-1} \in \mathbb{Q}(x_1, \ldots, x_n)$, a radical extension of $\mathbb{Q}(a_0, \ldots, a_{n-1})$ containing $\mathbb{Q}(x_1, \ldots, x_n)$ is also a radical extension of $\mathbb{Q}(x_1, \ldots, x_n)$. This is a good place to start proving something, because $\mathbb{Q}(x_1, \ldots, x_n)$ is symmetric with respect to $x_1, \ldots, x_n$ in the following sense (compare with Section 6.5*):

Proposition. *Any permutation σ of $x_1, \ldots, x_n$ extends to an automorphism of $\mathbb{Q}(x_1, \ldots, x_n)$.*

Proof. The extension of σ to $\mathbb{Q}(x_1, \ldots, x_n)$ is defined for each rational function $f(x_1, \ldots, x_n)$ of $x_1, \ldots, x_n$ by

$$\sigma f(x_1, \ldots, x_n) = f(\sigma x_1, \ldots, \sigma x_n).$$

This map is one-to-one and onto $\mathbb{Q}(x_1, \ldots, x_n)$ because it is inverted by σ^{-1}, and it preserves $+$ and $\times$ because if

$$f(x_1, \ldots, x_n) = g(x_1, \ldots, x_n) + h(x_1, \ldots, x_n)$$

then

$$f(\sigma x_1, \ldots, \sigma x_n) = g(\sigma x_1, \ldots, \sigma x_n) + h(\sigma x_1, \ldots, \sigma x_n)$$

and if

$$f(x_1, \ldots, x_n) = g(x_1, \ldots, x_n)h(x_1, \ldots, x_n)$$

then

$$f(\sigma x_1, \ldots, \sigma x_n) = g(\sigma x_1, \ldots, \sigma x_n)h(\sigma x_1, \ldots, \sigma x_n). \qquad \square$$

A radical extension E of $\mathbb{Q}(x_1, \ldots, x_n)$ is not necessarily symmetric in this sense. For example, $\mathbb{Q}(x_1, x_2, \sqrt{x_1})$ contains a square root of x_1, but not of x_2, hence there is no automorphism exchanging x_1 and x_2. However, we can restore symmetry by adjoining $\sqrt{x_2}$ as well. The obvious generalisation of this idea gives a way to "symmetrise" any radical extension E of $\mathbb{Q}(x_1, \ldots, x_n)$:

Theorem. *For any radical extension E of $\mathbb{Q}(x_1, \ldots, x_n)$ there is a radical extension $\overline{E} \supseteq E$ with automorphisms σ extending all permutations of $x_1, \ldots, x_n$.*

Proof. Since E can be built by successively adjoining p^{th} roots, it suffices to symmetrise $B(\alpha)$, where B is a symmetric radical extension of $\mathbb{Q}(x_1, \ldots, x_n)$ and $\alpha^p = b \in B$. We do this by adjoining a $\sigma\alpha$, defined by $(\sigma\alpha)^p = \sigma b$, for each permutation σ of $x_1, \ldots, x_n$, which we can assume to extend to an automorphism (also called σ) of B.

Then σ is further extended to an automorphism of $\overline{B} = B(\alpha, \sigma_1\alpha, \sigma_2\alpha, \ldots)$, where $\sigma_1, \sigma_2, \ldots$ denote all the nonidentity permutations of $x_1, \ldots, x_n$ (and their extensions to B), by setting

$$\sigma f(\alpha, \sigma_1\alpha, \sigma_2\alpha, \ldots, b_1, b_2, \ldots) = f(\sigma\alpha, \sigma\sigma_1\alpha, \sigma\sigma_2\alpha, \ldots, \sigma b_1, \sigma b_2, \ldots)$$

for each rational function f, where $b_1, b_2, \ldots \in B$. □

The reason for wanting an automorphism σ extending each permutation of $x_1, \ldots, x_n$ is that $a_0, \ldots, a_{n-1}$ are fixed by such permutations, hence so is every element of the field $\mathbb{Q}(a_0, \ldots, a_{n-1})$. The symmetry expressed by the theorem can therefore be restated as a property of Galois groups:

Corollary. *If E is a radical extension of $\mathbb{Q}(a_0, \ldots, a_{n-1})$ and $E \supseteq \mathbb{Q}(x_1, \ldots, x_n)$ then there is a radical extension $\overline{E} \supseteq E$ such that Gal$(\overline{E} : \mathbb{Q}(a_0, \ldots, a_{n-1}))$ includes automorphisms σ extending all permutations of $x_1, \ldots, x_n$.*

Proof. This is immediate from the theorem and the fact that a radical extension of $\mathbb{Q}(a_0, \ldots, a_{n-1})$ containing $\mathbb{Q}(x_1, \ldots, x_n)$ is also a radical extension of $\mathbb{Q}(x_1, \ldots, x_n)$. □

In short, a solution of the general n^{th} degree equation (*) by radicals yields a radical extension $\overline{E}$ of the coefficient field $\mathbb{Q}(a_0, \ldots, a_{n-1})$ which is "fully symmetric" in $x_1, \ldots, x_n$. On the other hand, the radical extensions we know from Section 8.1 seem to have rather limited symmetry, with each radical extending the Galois group by only a small step. This suggests a way to prove *nonexistence* of solutions by radicals, at least for $n \geq 5$, by showing that radical extensions have less than full symmetry in $x_1, \ldots, x_n$. In Section 8.3 we shall show that the symmetric extension $\overline{E}$ can be assumed to satisfy further conditions which imply that its Galois group has a special structure, called *solvability*, inherited from the sequence of adjoined radicals. Then in Section 8.4 we shall show that this structure precludes full symmetry in $x_1, \ldots, x_n$, when $n \geq 5$, thus completing the proof.

Exercise

8.2.1 Consider the extension $E = F(x_1)$ of $F = \mathbb{Q}(a_0, \ldots, a_4)$. Show

(i) $\sigma \in \text{Gal}(E : F)$ is determined by the value $\sigma(x_1)$.

(ii) $\sigma(x_1)$ is a root of $x^5 + a_4x^4 + \cdots + a_0 = 0$, hence $|\text{Gal}(E : F)| \leq 5$.

(iii) If some $x_i \neq x_1$ occurs as $\sigma(x_1)$ for a $\sigma \in \text{Gal}(E : F)$ then each $x_i \neq x_1$ occurs as a $\sigma(x_1)$.

(iv) If all $x_i \in E$ then $E = F(x_1, \ldots, x_5)$, hence not all $x_i \in E$. Why?

(v) $\text{Gal}(E : F) = \{1\}$.

8.3 Structure of Radical Extensions

If we revise the construction of $\overline{E}$ above to include all p^{th} roots of unity along with any other p^{th} root, then we can make the following assumptions about the adjoined radicals α_i in $\overline{E} = F(\alpha_1, \ldots, \alpha_k)$, where $F = \mathbb{Q}(a_0, \ldots, a_{n-1})$.

First, we can assume that each radical α_i adjoined is a pth root for some *prime* p. For example, instead of adjoining $\sqrt[6]{\alpha}$ we can adjoin first $\sqrt{\alpha} = \beta$, then $\sqrt[3]{\beta}$. Second, if α_i is a pth root but not a root of unity we can assume that $F(\alpha_1, \ldots, \alpha_{i-1})$ contains all p^{th} roots of unity, by placing the roots of unity first on the list of adjoined radicals. We can therefore assume:

The radical extension $\overline{E} = F(\alpha_1, \ldots, \alpha_k)$ of $F = \mathbb{Q}(a_0, \ldots, a_{n-1})$ *is the union of an ascending tower of fields*

$$F = F_0 \subseteq F_1 \subseteq \ldots \subseteq F_k = F(\alpha_1, \ldots, \alpha_k)$$

where each $F_i = F_{i-1}(\alpha_i)$, $\alpha_i^{p_i} \in F_{i-1}$, p_i *is prime, and* F_{i-1} *contains all* p_i^{th} *roots of unity if* α_i *is not a root of unity.*

Corresponding to this tower of fields we have a descending tower of groups

$$\text{Gal}(F_k : F_0) = G_0 \supseteq G_1 \supseteq \ldots \supseteq G_k = \text{Gal}(F_k : F_k) = \{\mathbf{1}\}$$

where $G_i = \text{Gal}(F_k : F_i) = \text{Gal}(F_k : F_{i-1}(\alpha_i))$ and $\mathbf{1}$ denotes the identity automorphism. The containments follow from the definition of $\text{Gal}(E : B)$, for any field $E \supseteq B$, as the group of automorphisms of E fixing each element of B. As B increases to E, $\text{Gal}(E : B)$ must decrease to $\{\mathbf{1}\}$. The important point is that the step from G_{i-1} to its subgroup G_i, reflecting the adjunction of the p_i^{th} root α_i to F, is "small" enough to be describable in group-theoretic terms: *G_i is a normal subgroup of G_{i-1}, and G_{i-1}/G_i is abelian,* as we shall now show.

To simplify notation further, we set

$$E = F_k, \ B = F_{i-1}, \ \alpha = \alpha_i, \ p = p_i,$$

then the theorem we want is:

Theorem. *If $E \supseteq B(\alpha) \supseteq B$ are fields with $\alpha^p \in B$ for some prime p, and if B contains all p^{th} roots of unity if α is not a root of unity, then Gal$(E : B(\alpha))$ is a normal subgroup of Gal$(E : B)$ and Gal$(E : B)/$Gal$(E : B(\alpha))$ is abelian.*

Proof. By the isomorphism theorem for groups, it suffices to find a homomorphism of $\text{Gal}(E : B)$, with kernel $\text{Gal}(E : B(\alpha))$, onto an abelian group. The obvious map with kernel $\text{Gal}(E : B(\alpha))$ is *restriction to* $B(\alpha)$, $|_{B(\alpha)}$, since by definition

$$\sigma \in \text{Gal}(E : B(\alpha)) \Leftrightarrow \sigma|_{B(\alpha)} \text{ is the identity map.}$$

The homomorphism property,

$$\sigma'\sigma|_{B(\alpha)} = \sigma'|_{B(\alpha)}\sigma|_{B(\alpha)} \text{ for all } \sigma', \sigma \in \text{Gal}(E : B),$$

is automatic provided $\sigma|_{B(\alpha)}(b) \in B(\alpha)$ for each $b \in B(\alpha)$, that is, provided $B(\alpha)$ is closed under each $\sigma \in \text{Gal}(E : B)$.

Since such a σ fixes B, $\sigma|_{B(\alpha)}$ is completely determined by the value $\sigma(\alpha)$. If α is a pth root of unity ζ then

$$(\sigma(\alpha))^p = \sigma(\alpha^p) = \sigma(\zeta^p) = \sigma(1) = 1,$$

hence $\sigma(\alpha) = \zeta^i = \alpha^i \in B(\alpha)$, since each pth root of unity is some ζ^i. If α is not a root of unity then

$$(\sigma(\alpha))^p = \sigma(\alpha^p) = \alpha^p \text{ since } \alpha^p \in B,$$

so $\sigma(\alpha) = \zeta^j\alpha$ for some p^{th} root of unity ζ. Since $\zeta \in B$ by hypothesis, we again have $\sigma(\alpha) \in B(\alpha)$. Thus $B(\alpha)$ is closed as required.

It now remains to check that the restricted automorphisms $\sigma|_B(\alpha)$ form an abelian group. If α is a root of unity then, as we have just seen, each $\sigma|_{B(\alpha)}$ is of the form σ_i, determined by the value $\sigma_i(\alpha) = \alpha^i$, and hence

$$\sigma_i\sigma_j(\alpha) = \sigma_i(\alpha^j) = \alpha^{ij} = \sigma_j\sigma_i(\alpha).$$

Likewise, if α is not a root of unity then each $\sigma|_{B(\alpha)}$ is of the form σ_i, determined by the value $\sigma_i(\alpha) = \zeta^i\alpha$, hence

$$\sigma_i\sigma_j(\alpha) = \sigma_i(\zeta^j\alpha) = \zeta^{i+j}\alpha = \sigma_j\sigma_i(\alpha)$$

since $\zeta \in B$ and therefore $\sigma(\zeta) = \zeta$. Hence in either case $|_{B(\alpha)}$ maps onto an abelian group. □

The property of $\text{Gal}(F(\alpha_1, \ldots, \alpha_k) : F)$ implied by this theorem, that it has subgroups

$$\text{Gal}(F(\alpha_1, \ldots, \alpha_k) : F) = G_0 \supseteq G_1 \supseteq \ldots \supseteq G_k = \{1\}$$

with each G_i normal in G_{i-1} and G_{i-1}/G_i abelian, is called *solvability* of $\text{Gal}(F(\alpha_1, \ldots, \alpha_k) : F)$.

Exercises

8.3.1 Let $E = F(x_1)$ be as in Exercise 8.2.1. Show that if E is contained in a radical extension of F then so is $F(x_i)$ for each i and hence so is $F(x_1, \ldots, x_5) = \mathbb{Q}(x_1, \ldots, x_5)$.

8.3.2 Illustrate the need to adjoin p^{th} roots of unity before other p^{th} roots with the subgroups $\text{Gal}(\mathbb{Q}(\sqrt[3]{2}, \zeta_3) : \mathbb{Q}(\zeta_3))$ and $\text{Gal}(\mathbb{Q}(\sqrt[3]{2}, \zeta_3) : \mathbb{Q}(\sqrt[3]{2}))$ of $\text{Gal}(\mathbb{Q}(\sqrt[3]{2}, \zeta_3) : \mathbb{Q})$.

8.4 Nonexistence of Solutions by Radicals when $n \geq 5$

As we know, this amounts to proving that a radical extension of $\mathbb{Q}(a_0, \ldots, a_{n-1})$ does not contain $x_1, \ldots, x_n$ or, equivalently, $\mathbb{Q}(x_1, \ldots, x_n)$. We have now reduced this problem to proving that the symmetry of the hypothetical extension $\overline{E}$ containing $x_1, \ldots, x_n$, given by the corollary to Theorem 8.2, is incompatible with the solvability of $\mathrm{Gal}(\overline{E} : \mathbb{Q}(a_0, \ldots, a_{n-1}))$, given by Theorem 8.3. Our proof looks only at the effect of the hypothetical automorphisms of $\overline{E}$ on $x_1, \ldots, x_n$, and hence it is really about the *symmetric group* S_n of all permutations of $x_1, \ldots, x_n$. In fact, we are adapting a standard proof that S_n is not a solvable group, given by Milgram in his appendix to Artin [1942]. Apart from some basic group theory, this proof involves only the notion of a 3-*cycle* (x, y, z), which is a permutation θ of objects $x, y, z, \ldots$ such that $\theta(x) = y$, $\theta(y) = z$, $\theta(z) = x$ and all other objects are fixed.

Theorem. *When $n \geq 5$, a radical extension of $\mathbb{Q}(a_0, \ldots, a_{n-1})$ does not contain $\mathbb{Q}(x_1, \ldots, x_n)$.*

Proof. Suppose on the contrary that E is a radical extension of $\mathbb{Q}(a_0, \ldots, a_{n-1})$ containing $\mathbb{Q}(x_1, \ldots, x_n)$. Then E is also a radical extension of $\mathbb{Q}(x_1, \ldots, x_n)$, and by the corollary to Theorem 8.2 there is a radical extension $\overline{E} \supseteq E$ such that $G_0 = \mathrm{Gal}(\overline{E} : \mathbb{Q}(a_0, \ldots, a_{n-1}))$ includes automorphisms σ extending all permutations of $x_1, \ldots, x_n$.

By Theorem 8.3, G_0 has a decomposition

$$G_0 \supseteq G_1 \supseteq \ldots \supseteq G_k = \{1\}$$

where each G_{i+1} is a normal subgroup of G_i and G_{i-1}/G_i is abelian. We now show that this contradicts the existence of the automorphisms σ.

Since G_{i-1}/G_i is abelian, G_i is the kernel of a homomorphism of G_{i-1} onto an abelian group, and therefore (compare with Section 7.9, Example 3)

$$\sigma, \tau \in G_{i-1} \Rightarrow \sigma^{-1}\tau^{-1}\sigma\tau \in G_i.$$

We use this fact to prove by induction on i that, if $n \geq 5$, each G_i contains automorphisms σ extending all 3-cycles (x_a, x_b, x_c). This is true for G_0 by hypothesis, and when $n \geq 5$ the property persists from G_{i-1} to G_i because

$$(x_a, x_b, x_c) = (x_d, x_a, x_c)^{-1}(x_c, x_e, x_b)^{-1}(x_d, x_a, x_c)(x_c, x_e, x_b)$$

where a, b, c, d, e are distinct. Thus if there are at least five indeterminates x_j, there are σ in each G_i which extend arbitrary 3-cycles (x_a, x_b, x_c), and this means in particular that $G_k \neq \{1\}$.

This contradiction shows that $\mathbb{Q}(x_1, \ldots, x_n)$ is not contained in any radical extension of $\mathbb{Q}(a_0, \ldots, a_{n-1})$ when $n \geq 5$. □

Corollary. *The general equation of degree $n \geq 5$ is not solvable by radicals.*

Proof. Immediate from the theorem above and the formulation of solution by radicals in Section 8.2. □

Remarks. S_5 does have one proper normal subgroup, A_5, and the quotient S_5/A_5 is the abelian group C_2. However, the chain $S_5 \supseteq A_5$ cannot be continued because A_5 has *no* proper normal subgroup. This remarkable result can also be proved by considering 3-cycles (see Exercises 8.4.3 and 8.4.4), though with rather more trouble. It can also be shown that any chain of normal subgroups with abelian quotients must begin with $S_5 \supseteq A_5$, hence A_5 is the real obstruction to solvability of S_5.

It follows from the solution of the general cubic and quartic equations by radicals (Sections 1.6 and 1.7), and Section 8.3, that S_3 and S_4 are solvable. This can also be proved directly, of course (Exercise 8.4.2), for example by studying the subgroups of S_3 and S_4 we have already found (Section 7.8).

Exercises

8.4.1 Deduce from Exercises 8.2.1, 8.3.1 and the theorem just given that the group $\mathrm{Gal}(E : F)$ can be solvable, even equal to $\{1\}$, without E being contained in a radical extension of F.

8.4.2 Find decompositions of S_3 and S_4 into subgroups that demonstrate their solvability.

8.4.3 Suppose that $H \neq \{1\}$ is a normal subgroup of A_5, and recall (from Section 7.5.2) that A_5 consists of all products of 3-cycles. Show

(i) Any $\alpha \in H$ is either a 5-cycle, a 3-cycle or a product of two disjoint 2-cycles (transpositions).

(ii) If $\alpha = (a,b,c,d,e) \in H$ and $\beta = (a,b,c)$, then $\beta\alpha\beta^{-1}\alpha^{-1} \in H$ and $\beta\alpha\beta^{-1}\alpha^{-1}$ is a 3-cycle.

(iii) If $\alpha = (a,b)(c,d) \in H$ is a product of disjoint 2-cycles, and $\beta = (a,b,e)$, then $\beta\alpha\beta^{-1}\alpha^{-1} \in H$ and $\beta\alpha\beta^{-1}\alpha^{-1}$ is a 3-cycle.

(iv) Hence H includes a 3-cycle.

8.4.4 Deduce from Exercise 8.4.3 that if H is a nontrivial normal subgroup of A_5 then H includes all 3-cycles and hence is all of A_5.

8.5* Quintics with Integer Coefficients

It is not at all obvious that unsolvability of the general equation of degree $n \geq 5$ implies unsolvability of any particular equation with numerical coefficients. The problem is that the roots $\alpha_1, \ldots, \alpha_n$ of a particular polynomial $p(x)$, from $\mathbb{Q}[x]$ say, do not necessarily have the symmetry of the indeterminates $x_1, \ldots, x_n$, which are fully symmetric virtually by definition. Recall from Section 8.2 that "full symmetry" means that each permutation of $x_1, \ldots, x_n$ extends to an automorphism of $\mathbb{Q}(x_1, \ldots, x_n)$. The root field $\mathbb{Q}(\alpha_1, \ldots, \alpha_n)$ of a polynomial $p(x)$ does not necessarily admit this many automorphisms, but it does admit a promising set of automorphisms when $p(x)$ is irreducible:

Theorem. *If $p(x) \in \mathbb{Q}[x]$ is irreducible, with roots $\alpha_1, \ldots, \alpha_n$, then for each pair α_i, α_j there is an automorphism σ_{ij} of $\mathbb{Q}(\alpha_1, \ldots, \alpha_n)$ with $\sigma_{ij}(\alpha_j) = \alpha_j$.*

Proof. We know from the conjugation theorem of Section 6.2 that there is an isomorphism σ_{ij} of $\mathbb{Q}(\alpha_i)$ onto $\mathbb{Q}(\alpha_j)$ with $\sigma_{ij}(\alpha_j) = \alpha_j$. It then follows from the isomorphism extension theorem of Section 6.3 that σ_{ij} extends to $\mathbb{Q}(\alpha_1, \ldots, \alpha_n)$, by adjoining the first root α_k not in $\mathbb{Q}(\alpha_i)$, then the first root α_m not in $\mathbb{Q}(\alpha_i, \alpha_k)$, etc..

Moreover, the values $\sigma_{ij}(\alpha_1), \ldots, \sigma_{ij}(\alpha_n)$ must be distinct roots of $p(x)$, hence they are just the roots $\alpha_1, \ldots, \alpha_n$ again, possibly in a different order. This means that the extended isomorphism σ_{ij} is onto $\mathbb{Q}(\alpha_1, \ldots, \alpha_n)$, and hence it is an automorphism. □

A group G of permutations of $\alpha_1, \ldots, \alpha_n$ with the property described in the theorem – that G includes elements sending any α_i to any α_j – is called *transitive.* Thus the theorem says that the root field of an irreducible polynomial has a transitive group of automorphisms. Unfortunately, this is not enough to guarantee automorphisms extending all permutations of $\alpha_1, \ldots, \alpha_n$. For example, the root field $\mathbb{Q}(\zeta_5)$ of $x^4 + x^3 + x^2 + x + 1$ has automorphisms sending each root ζ_5^i to each other root ζ_5^j $(1 \leq i, j \leq 4)$, but there are only four automorphisms altogether (namely, $\sigma_k(\zeta_5) = \zeta_5^k$ for $k = 1, 2, 3, 4$).

It turns out that this shortage is avoided whenever the automorphisms include a *transposition* – an automorphism that exchanges two roots α_i, α_j and leaves the others fixed. We shall prove this result in the next section. Meanwhile, we construct an example of a *quintic* (fifth degree) irreducible polynomial whose root field actually admits a transposition.

Example. The polynomial $x^5 - 4x + 2$ is irreducible, and complex conjugation is an automorphism of its root field which exchanges two roots and leaves three roots fixed.

Eisenstein's irreducibility criterion with $p = 2$ shows that $x^5 - 4x + 2$ is irreducible. Since the derivative $5x^4 - 4$ has two real roots, the real graph of $y = x^5 - 4x + 2$ has two turning points. Then the sign changes of $x^5 - 4x + 2$ between $x = -2, 0, 1$ and 2 show that it has three real roots $\alpha_1, \alpha_2, \alpha_3$. The other two roots β_1, β_2, which exist by the factor theorem and the fundamental theorem of algebra (Section 3.8), must therefore be nonreal. They must also be complex conjugates of each other, because if $0 = \beta^5 - 4\beta + 2$ then $0 = \overline{\beta}^5 - 4\overline{\beta} + 2$ by applying complex conjugation to both sides.

Since complex conjugation fixes the real roots $\alpha_1, \alpha_2, \alpha_3$ and exchanges β_1, β_2 it is a permutation of $\{\alpha_1, \alpha_2, \alpha_3, \beta_1, \beta_2\}$, and hence maps $\mathbb{Q}(\alpha_1, \alpha_2, \alpha_3, \beta_1, \beta_2)$ one-to-one onto itself. It also preserves $+$ and $\times$, of course, and therefore is an automorphism. □

Exercises

8.5.1 Find some other irreducible quintic polynomials with three real roots, hence with root fields admitting transpositions.

8.5.2 Find some irreducible quartic polynomials whose root fields admit transpositions.

8.5.3 Show that the root field of $x^3 - 2$ is $\mathbb{Q}(\sqrt[3]{2}, \zeta_3)$, and that it admits a transposition when viewed as $\mathbb{Q}(\sqrt[3]{2}, \zeta_3\sqrt[3]{2}, \zeta_3^2\sqrt[3]{2})$.

8.6* Unsolvable Quintic Equations with Integer Coefficients

Now that we have the quintic polynomial $x^5 - 4x + 2$ whose root field admits a transposition, we can prove that the equation $x^5 - 4x + 2 = 0$ is not solvable by radicals. It suffices to prove the following lemma about permutation groups.

Lemma. *If p is prime and G is a transitive subgroup of S_p which includes a transposition, then $G = S_p$.*

Proof. View S_p as the group of permutations of the set $\{1, 2, \ldots, p\}$ and define, for each $i \in \{1, 2, \ldots, p\}$,

$$\text{class of } i = \{j : (i, j) \in G\},$$

where (i, j) denotes the transposition that exchanges i and j. Any two of these classes are either identical or disjoint. If, say, k is in the classes of m and n then $(m, k) \in G$, $(n, k) \in G$ and hence $(m, n) = (m, k) \cdot (n, k) \in G$, so the two classes are the same.

Moreover, any two of these classes are the same size. In fact, if σ_k is a permutation in G such that $\sigma_k(1) = k$ (such a permutation exists by the transitivity of G) then σ_k maps the class of 1 one-to-one onto the class of k. The key to this fact is the formula

$$\sigma_k \cdot (1, i) \cdot \sigma_k^{-1} = (k, \sigma_k(i)),$$

which is easily checked when one remembers to read the product of three terms from right to left. It follows from the formula that

$$\begin{aligned} i \in \text{class of } 1 &\Rightarrow (1, i) \in G \\ &\Rightarrow \sigma_k \cdot (1, i) \cdot \sigma_k^{-1} \in G \\ &\Rightarrow (k, \sigma_k(i)) \in G \\ &\Rightarrow \sigma_k(i) \in \text{class of } k. \end{aligned}$$

A similar argument shows that

$$i \in \text{class of } k \Rightarrow \sigma_k^{-1}(i) \in \text{class of } 1,$$

hence σ_k is the one-to-one correspondence claimed.

It follows that the size of each class divides p, the size of $\{1, 2, \ldots, p\}$, and hence is either 1 or p since p is prime. If we now assume, without loss of generality, that (1,2) is a transposition in G, then the class of 1 has at least two elements, 1 and 2. Hence there is only one class – of p elements – and this means that G includes each transposition (i, j).

Finally, we recall that every permutation is a product of transpositions. This can be proved, for example, by expressing each permutation σ as the product of disjoint cycles – first $(1, \sigma(1), \sigma^2(1), \ldots)$, then $(i, \sigma(i), \sigma^2(i), \ldots)$ where i is the least number not in the first cycle, etc. – then expressing an arbitrary m-cycle as a product of transpositions, namely

$$(i_1, i_2, \ldots, i_m) = (i_1, i_m)(i_1, i_{m-1}) \cdots (i_1, i_2). \qquad \square$$

With this lemma in hand, the proof that $x^5 - 4x + 2$ is not solvable by radicals is now just a recapitulation of the main results of the last four sections.

Theorem. *The equation $x^5 - 4x + 2 = 0$ is not solvable by radicals.*

Proof. This equation is the example of Section 8.5*, where it was found to have three real roots $\alpha_1, \alpha_2, \alpha_3$ and two nonreal roots β_1, β_2 which are complex conjugates. It follows from Theorem 8.5* that the root field $\mathbb{Q}(\alpha_1, \alpha_2, \alpha_3, \beta_1, \beta_2)$ has a transitive group of automorphisms, and from the example in Section 8.5* that this group includes a transposition. Hence it follows from the lemma just given that there are automorphisms of $\mathbb{Q}(\alpha_1, \alpha_2, \alpha_3, \beta_1, \beta_2)$ extending all permutations of the roots.

But then it follows from the proof of Theorem 8.4 that $\mathbb{Q}(\alpha_1, \alpha_2, \alpha_3, \beta_1, \beta_2)$ is contained in no radical extension of $\mathbb{Q}$. In particular, the roots $\alpha_1, \alpha_2, \alpha_3, \beta_1, \beta_2$ are not all expressible by radicals. $\square$

Exercises

8.6.1 Show that the dihedral group D_4 is isomorphic to a transitive subgroup of S_4 which includes a transposition.

8.6.2 Show that the field $\mathbb{Q}(\sqrt[4]{2}, i)$ (whose automorphism group is D_4 by Exercise 7.4.2) is the root field of an irreducible quartic polynomial.

8.7* Primitive Roots

The necessary condition for solvability by radicals, namely solvability of the Galois group, can be tightened up considerably. The abelian quotient groups G_{i-1}/G_i can be taken to be cyclic. We shall prove this in Section 8.8, but before doing so it is of interest to see why this is true of the particular groups constructed in Section 8.3. Each group G_{i-1}/G_i constructed there is either a subgroup of $\mathbb{Z}/p\mathbb{Z}$ under addition mod p, that is, a subgroup of the cyclic group C_p, or of $\mathbb{Z}/p\mathbb{Z} - \{0\}$ under multiplication mod p.

Now a subgroup of any cyclic group C_m can be seen to be cyclic by the following argument. View C_m as $\{0, 1, 2, \ldots, m-1\}$ under addition mod m, and let d be the least nonzero member of a subgroup H. Then $0, d, 2d, \ldots \in H$ by closure under addition. If H includes any element $nd + r$ with $0 < |r| < |d|$ we also have $r \in H$ by closure under subtraction. This contradicts the minimality of d, hence H is the cyclic group $\{0, d, 2d, \ldots\}$.

In view of this fact, to show that G_{i-1}/G_i is cyclic in Section 8.3 it only remains to show that $(\mathbb{Z}/p\mathbb{Z})^\times$ is cyclic. This is a special case of the next theorem, that $F - \{0\}$ is a cyclic group under multiplication for any finite field F. The proof of this theorem depends on the following lemma about abelian groups. The crucial concept in its proof is the *order* of an element u in an abelian group, the least m such that $mu = u + u \cdots + u$ (m times) equals 0.

Lemma. *If elements* a,b *of an abelian group are such that* $\text{order}(a) = m$, $\text{order}(b) = n$ *and* $\gcd(m, n) = 1$ *then* $\text{order}(a + b) = mn$.

Proof. Certainly $mn(a + b) = mna + mnb = 0$, hence

$$\begin{aligned} &\text{order}(a + b) \neq mn \\ \Rightarrow\ &l(a + b) = la + lb = 0 \quad \text{for some} \quad l < mn \\ \Rightarrow\ &ra + sb = 0 \quad \text{for some} \quad 0 \leq r < m,\ 0 \leq s < n \end{aligned}$$

where

$$\begin{aligned} r &= \text{remainder of } l \text{ on division by } m, \\ s &= \text{remainder of } l \text{ on division by } n, \end{aligned}$$

are not both zero since $l < mn = \text{lcm}(m, n)$, and in fact they are both nonzero otherwise $\text{order}(a) = r < m$ or $\text{order}(b) = s < n$.

$$\begin{aligned} &\Rightarrow ra = -sb \\ &\Rightarrow rna = -snb = 0 \quad \text{since} \quad nb = 0 \\ &\Rightarrow \text{order}(a) = m | rn \\ &\Rightarrow m | r \quad \text{since} \quad \gcd(m, n) = 1 \end{aligned}$$

and this is a contradiction since $0 < r < m$. □

Corollary. *If* a *is an element of maximal order in a finite abelian group* A, *then* $\text{order}(b) | \text{order}(a)$ *for any other element* $b \in A$.

Proof. Suppose on the contrary that $b \in A$ is such that $\text{order}(b) \nmid \text{order}(a)$, so that there is some prime p with

$$\text{order}(a) = p^i q,\ \text{order}(b) = p^j r \quad \text{where} \quad p \nmid q, r \quad \text{and} \quad i < j.$$

Then $b' = rb$ has order p^j which is relatively prime to the order q of $a' = p^i a$, and hence $a' + b'$ has order $p^j q > \text{order}(a)$, contrary to the maximality of $\text{order}(a)$. □

Theorem. *If F is a finite field then $F - \{0\}$ is a cyclic group under multiplication.*

Proof. $F - \{0\}$ is an abelian group because it is closed under $\times$ and $\div$, and $\times$ is commutative. Let $\alpha \in F - \{0\}$ be an element of maximal order, d. This means that α is a root of the polynomial $x^d - 1$, and the powers $1, \alpha, \alpha^2, \ldots, \alpha^{d-1}$ are distinct. These powers are also roots of $x^d - 1$, so $x^d - 1$ has d distinct roots.

If $F - \{0\} \neq \{1, \alpha, \ldots, \alpha^{d-1}\}$, suppose β is another element of $F - \{0\}$. The order of β divides the order of α by the corollary, and hence β also satisfies $x^d - 1 = 0$. This contradicts the fact that a polynomial of degree d over a field F cannot have more than d roots in F (Section 4.2). □

The classical case of this theorem, conjectured by Euler and proved by Gauss [1801], is where $F = \mathbb{Z}/p\mathbb{Z}$. In this case the element α is called a *primitive root* mod p and the order of α is $p - 1$. The indirect way its existence is proved leads one to suspect that questions about primitive roots may be difficult. Indeed they are. For example, it is not known whether 2, or any other particular number, is a primitive root for infinitely many p. Gauss [1801], article 315, was interested in the p for which 10 is a primitive root, because these are precisely the p for which the decimal expansion of $1/p$ has period $p - 1$ (see Exercises below). It is likewise unknown whether there are infinitely many such p.

Exercises

8.7.1 Show that the period of the decimal expansion of $1/n$ is the least exponent e such that $10^e \equiv 1 \pmod{n}$.

8.7.2 Deduce from Exercise 8.7.1 that the decimal expansion of $1/n$ has period $n - 1 \Leftrightarrow n$ is prime and 10 is a primitive root mod n.

8.7.3 Is it always true that $\text{order}(a + b) = \text{lcm}(\text{order}(a), \text{order}(b))$ in a finite abelian group?

8.8 Finite Abelian Groups

The following lemma is the key to a tighter definition of a solvable group. By breaking down a finite abelian group into cyclic "steps," it points the way to a breakdown of any solvable group into cyclic "steps."

Lemma. *Each finite abelian group A admits a decomposition into subgroups*

$$A = A_0 \supseteq A_1 \supseteq \ldots \supseteq A_l = \{1\},$$

such that each A_{j-1}/A_j is cyclic of prime order.

Proof. If A_0 is not already cyclic of prime order, let A_1 be a maximal proper subgroup of A_0. That is, $A_1 \neq A_0$, and there is no group B strictly between A_1 and A_0. Since A_0 is abelian, any subgroup is normal, hence we have the coset homomorphism

$$\psi : A_0 \to A_0/A_1$$

with kernel A_1. I claim that A_0/A_1 is cyclic of prime order.

If it is not cyclic, then any $g \neq 1$ in A_0/A_1 generates a nontrivial cyclic subgroup $G = \{1, g, g^2, \ldots\}$, which is not all of A_0/A_1. But then $B = \psi^{-1}(G)$ is a set strictly between A_1 and A_0, and it is easily checked that B is closed under products and inverses, hence a group. This contradicts the choice of A_1, so A_0/A_1 is in fact cyclic. It follows similarly that A_0/A_1 has prime order, because if A_0/A_1 has order pq and generator g then the powers of g^p form a proper nontrivial subgroup G of A_0/A_1.

Thus we have found a subgroup A_1 of A_0 with the required property. We can now repeat the argument with A_1 in place of A_0, and obtain A_2, A_3, etc., also with the required properties. Since each A_j is strictly smaller than A_{j-1}, in a finite number of steps we reach $A_l = \{1\}$. □

Now if $G_{i-1} \supseteq G_i$ are groups with G_i normal in G_{i-1} and $G_{i-1}/G_i = A$ abelian, we can decompose the space between G_{i-1} and G_i into cyclic "steps" by "pulling back" the decomposition of A given by the lemma. To be precise, we have the following:

Theorem. *If a finite group G admits a decomposition into subgroups*

$$G = G_0 \supseteq G_1 \supseteq \ldots \supseteq G_k = \{1\}$$

with each G_i normal in G_{i-1} and G_{i-1}/G_i abelian, then it admits a finer decomposition

$$G_{i-1} = H_{i0} \supseteq H_{i1} \supseteq \ldots \supseteq H_{il} = G_i$$

with each H_{ij} normal in $H_{i,j-1}$ and $H_{i,j-1}/H_{ij}$ cyclic of prime order.

Proof. Let A be the finite abelian group G_{i-1}/G_i and let $\chi : G_{i-1} \to A$ be the coset homomorphism with kernel G_i. Consider the decomposition of A into subgroups

$$A = A_0 \supseteq A_1 \supseteq \ldots \supseteq A_l = \{1\}$$

with each A_{j-1}/A_j cyclic of prime order, given by the lemma.

If we let $H_{ij} = \chi^{-1}(A_j)$ then each H_{ij} is a group and

$$\chi^{-1}(A) = G_{i-1} = H_{i0} \supseteq H_{i1} \supseteq \ldots \supseteq H_{il} = G_i = \chi^{-1}(\{1\})$$

because

$$\chi(G_{i-1}) = A_0 \supseteq A_1 \supseteq \ldots \supseteq A_l = \{1\} = \chi(G_i).$$

Thus it suffices to show that H_{ij} is a normal subgroup of $H_{i,j-1}$ with $H_{i,j-1}/H_{ij}$ cyclic of prime order. Equivalently: find a homomorphism of $H_{i,j-1}$ onto a cyclic group of prime order, with kernel H_{ij}.

We already have the homomorphism χ of $H_{i,j-1}$ onto A_{j-1}, and a homomorphism ψ of A_{j-1} onto A_{j-1}/A_j with kernel A_j. Hence $\psi\chi$ is a homomorphism of $H_{i,j-1}$ onto the cyclic group A_{j-1}/A_j of prime order, and its kernel $\chi^{-1}(A_j)$ is H_{ij}, as required. □

Corollary. *A finite group G is solvable $\Leftrightarrow$ G admits a decomposition into subgroups*

$$G = H_0 \supseteq H_1 \supseteq \ldots \supseteq H_m = \{1\}$$

with each H_n normal in H_{n-1} and H_{n-1}/H_n cyclic of prime order.

Proof. ($\Rightarrow$) is immediate from the theorem.

($\Leftarrow$) is immediate from the definition of solvable group in Section 8.3, since cyclic groups are abelian. □

The narrower definition of solvability given by this corollary will be exploited in the next chapter. There we will show that an equation is solvable by radicals if its root field has a solvable Galois group G, and in fact the cyclic "steps" of prime order p in the decomposition of G correspond to p^{th} roots. In particular, it will follow from the breakdown of abelian groups into cyclic "steps" that any equation with abelian Galois group is solvable by radicals. A theorem equivalent to this was discovered by Abel [1829]. In honour of this discovery, Kronecker introduced the term "abelian fields" for fields with abelian Galois group over $\mathbb{Q}$, from which we derive the term "abelian groups." Kronecker also made the remarkable discovery that every abelian field is contained in a cyclotomic field (Kronecker [1853]; the proof was completed by Weber [1886]).

Exercise

8.8.1 Give an example of abelian groups A_0, A_1 with $A_0/A_1 \cong C_m$ and $A_1 \cong C_n$ but $A_0 \ncong C_m \times C_n$. (Thus it does not follow from the lemma – at least, not obviously – that any finite abelian group is the direct product of cyclic groups.)

8.9 Discussion

Galois theory has had an enormous influence on the development of algebra, and also on the way it is taught. As was mentioned in Section 7.10, most of the fundamental concepts of group theory are prerequisites for Galois theory, and one can now see why. The concept of normal subgroup is crucial to an understanding of radical extensions, and also for the understanding of S_n that shows the impossibility of solvability by radicals. The very term "solvable group" gives away its origin in the theory of equations. The word "abelian," too, comes from a class of equations studied by Abel [1829], as we have just seen.

However, it seems to be one of the laws of mathematical history that if a concept *can* be detached from its origins, it will be. This is particularly true of the group concept, for which the definition is simple and for which examples are more readily available in number theory and geometry than the theory of equations. No doubt this is why most algebra texts treat groups before they treat fields. Even the present treatment is not very faithful to the history of the subject, because some modern concepts are simply too efficient to do without. It is therefore worth saying a few words about the history of Galois theory and its influence on the rest of algebra. More detail can be found in the books of Edwards [1984] and Tignol [1988], and in the article of Kiernan [1971]. Edwards is also worth reading for his translation of Galois' fundamental memoir [1831].

The first reasonably complete proof of the unsolvability of the general quintic equation by radicals was given by Abel [1826], after incomplete attempts by Ruffini [1799] and [1813]. Ruffini and Abel observed the special properties of permutations of five things underlying unsolvability of the quintic, but did not express them in group theoretic language.

The great contribution of Galois was to identify the group concept and to see how it clarifies the situation. It enabled him to analyse the problem of solution by radicals in complete generality and to find a criterion for solvability which has unsolvability of the general quintic as just one of many corollaries. Galois [1831] found that a solvable equation has a solvable group by much the same route we used in Section 8.3. He broke down the process of adjoining radicals to adjunctions of pth roots, with p^{th} roots of unity adjoined before other p^{th} roots, and showed that the corresponding group theoretic step was passage to a normal subgroup with cyclic quotient. However, he used the solvable group criterion only as a stepping stone to one he thought would appeal more to his contemporaries – that an equation is solvable by radicals if and only if any two of its roots rationally determine all the others. (See Tignol [1988], p.381 for a modern proof of this theorem.)

It is understandable that Galois felt the group concept was too foreign, and the concept of solvable group perhaps too complicated, to provide an attractive criterion for solvability of equations. In the algebra of his time, Galois' root criterion was both elegant and surprising. Unfortunately, his proof was too sketchy to be understood by his referees, and it was not published in Galois' brief lifetime (he died after a duel in 1832, aged 20). In the 1840s Liouville studied the proof and became convinced it was correct, so he published Galois' paper in 1846. This brought the group concept to the attention of other mathematicians, and over the next two decades it was assimilated to the point where Jordan could write his *Traité des substitutions et des équations algébriques* [1870], a book inspired by Galois theory in which group theory takes over almost completely. In particular, the *Traité* states solvability of *groups* as the criterion for solvability of equations, for the first time, and begins the investigation of solvable groups in their own right. The *Traité* also stresses the concept of *simple* group, which was important in early proofs of unsolvability by radicals, and has since become a topic of spectacular interest in the theory of finite groups.

A group is called *simple* if its only quotients are itself and $\{\mathbf{1}\}$; equivalently, if its only normal subgroups are itself and $\{\mathbf{1}\}$. Perhaps the best way to understand the concept is to think of a simple group as one that admits no "simplification," that is, no homomorphism onto a smaller nontrivial group. "Simple" does not mean uncomplicated! About the only uncomplicated examples are the cyclic groups C_p for prime p. The first interesting example of a simple group is A_5, and its simplicity was in fact used in early proofs of the unsolvability of the quintic. Apparently a direct proof that S_5 is not solvable was not noticed until later.

Many algebra textbooks are still under the influence of Jordan's *Traité* in the sense that they prepare for Galois theory by an extensive study of solvable and simple groups. While this is unnecessary, it is perhaps excusable in the light of modern discoveries about simple groups.

Apart from a few infinite families, such as the A_n for $n \geq 5$, finite simple groups are extraordinarily rare and hard to find. A complete list was obtained only in the last decade, and some steps in the 10,000+ page proof still remain to be published. One of the key "details" is the 250 page proof of Feit and Thompson [1963] that every group with an odd number of elements is solvable. One could say that the classification of finite simple groups is the deepest (and most mysterious?) theorem in mathematics to date.

As we have mentioned in Section 7.10, Galois groups were viewed as groups of permutations of the roots of equations until the 1870s, because the field concept did not emerge until that time. Certainly, there is not much difference between the group of automorphisms of a field E and its restriction to a set of basis elements of E. This is particularly noticeable in Sections 8.4, 8.5*, 8.6*, where we are interested *only* in the way an automorphism permutes certain elements of the field, and we use the word "transposition" indifferently for both permutations and automorphisms. The advantage of the field concept is its independence of a particular equation or a particular basis, which enables one to see its invariant properties, such as dimension. As Dedekind realised, the property of dimension alone is strong enough to settle questions like the constructibility of $\sqrt[3]{2}$ (Section 5.6).

Dedekind developed the field concept in his supplements to Dirichlet's *Vorlesungen* [1871-1893]. The 1893 edition (§166) contains the definition of Galois groups as automorphism groups of fields. It is interesting that Dedekind pays a kind of tribute to the original concept of Galois group by calling automorphisms "permutations." In fact, he even calls an isomorphism of a field onto a different field a "permutation."

However, Dedekind's definition is not simply a translation of the permutation group concept into the language of fields. In the field setting, new and useful properties of the Galois group become apparent. For example, Dedekind [1893], §165, discovered that, under reasonable conditions, the size of $\mathrm{Gal}(E : F)$ is the same as the dimension $(E : F)$ of E over F. This result prompted a new approach to Galois theory, emphasising connections with linear algebra and playing down the role of polynomials. This approach is due mainly to Emil Artin, though in Artin's book [1942] the polynomials are still present to a greater extent than is theoretically necessary. In particular, Artin uses them to prove his so-called "fundamental theorem of Galois theory," although a polynomial-free proof is possible (see, for example, the appendix to Tignol [1988]). Since our interest is very much in polynomials, we have not adopted Artin's approach in this book. Our proof of the fundamental theorem of Galois theory is in Section 9.3. Nothing like this theorem appears in the work of Galois, but it does help to clarify the converse direction of his solvability criterion, which we have yet to prove.

9 Galois Theory of Solvability

9.1 The Theorem of the Primitive Element

The unsolvability of general polynomial equations of degree ≥ 5 leaves us with very little to say about *solvability* of general equations, that is, equations with indeterminate coefficients. The general linear, quadratic, cubic and quartic equations are solvable – and that's it. The investigation of solvability is much more fruitful in the domain of equations with numerical coefficients, where there are solvable equations of arbitrarily high degree. For this reason, **all fields in this chapter are assumed to be number fields, that is, subfields of $\mathbb{C}$, unless there is an explicit statement to the contrary**. We shall be particularly interested in number fields of finite degree over $\mathbb{Q}$, which we know from Chapter 5 to be of the form $\mathbb{Q}(\alpha_1, \ldots, \alpha_k)$ where $\alpha_1, \ldots, \alpha_k$ are algebraic numbers.

As a first step in the study of such fields we show that they can be simplified to the form $\mathbb{Q}(\alpha)$. More generally, we show that if E, F are fields with $(E : F)$ finite then $E = F(\alpha)$ for some $\alpha \in E$ called a *primitive element*. This ties up a loose end from Section 5.3 where we called $(E : F) = n$ the *degree* of E over F whether or not there was an element $\alpha \in E$ of degree n over F with $F(\alpha) = E$. The existence of such an element for number fields goes back to Lagrange, Abel and Galois and is called the *theorem of the primitive element*.

The proof depends on the lemma below, which is most easily proved with the help of calculus. It is possible to get around the calculus by checking a certain algebraic identity (see Exercise 9.1.1 and also compare with Section 4.9*), but as long as we are dealing with number fields this is unnecessary extra work.

Recall from the factor theorem of Sections 3.8 and 4.2 that each root $\alpha \in F$ of a polynomial $f(x) \in F[x]$ corresponds to a factor $x - \alpha$ of $f(x)$. We say that $f(x)$ has a *multiple root* α if the factor $x - \alpha$ is repeated.

Lemma. *If $f \in F[x]$ is irreducible over the field $F \subseteq \mathbb{C}$ then f has no multiple root in $\mathbb{C}$.*

Proof. Suppose on the contrary that over $\mathbb{C}$

$$f(x) = (x - \alpha)^2 g(x).$$

Then the derivative

$$f'(x) = (x - \alpha)^2 g'(x) + 2(x - \alpha)g(x) = (x - \alpha)((x - \alpha)g'(x) + 2g(x)),$$

and hence $f(x), f'(x)$ have a common divisor $x - \alpha$. Thus $\gcd(f(x), f'(x))$ has degree ≥ 1, and of course the gcd also has degree $<$ degree(f), since degree(f') = degree$(f) - 1$.

Now if $f(x) = \sum_i a_i x^i \in F[x]$ then $f'(x) = \sum_i i a_i x^{i-1} \in F[x]$ also, hence we can find $\gcd(f(x), f'(x))$ *in* $F[x]$ by the Euclidean algorithm. Since the gcd is a proper divisor of f, this contradicts the irreducibility of f. □

Theorem of the Primitive Element. *If $E, F \subseteq \mathbb{C}$ are fields with $(E:F)$ finite then $E = F(\alpha)$ for some $\alpha \in E$.*

Proof. Suppose $(E:F) = n$. Then by Section 5.4 we have $E = F(\alpha_1, \ldots, \alpha_n)$ where each $\alpha_i \in E$ is a root of some $f_i(x) \in F[x]$ of degree $\leq n$. It obviously suffices to show that if $\beta, \gamma \in E$ then $F(\beta, \gamma) = F(\alpha)$ for some $\alpha \in E$.

Since F is a number field it is infinite and hence there is a $c \in F$ such that

$$\beta + c\gamma \neq \beta_i + c\gamma_j$$

for all roots β_i of $g(x)$ and γ_j of $h(x)$, where $g(x), h(x) \in F[x]$ are irreducible polynomials satisfied by β, γ respectively (see Corollary 5.4). Namely, take c unequal to any of the solutions $y = (\beta - \beta_i)/(\gamma_j - \gamma)$ of $\beta + y\gamma = \beta_i + y\gamma_i$. We let $\alpha = \beta + c\gamma$.

Now the polynomials $h(x)$ and $g(\alpha - cx)$ in $F(\alpha)[x]$ have γ as a common root, and it is their only common root by choice of c. This has the following implications for the irreducible monic polynomial for γ, $f(x) \in F(\alpha)[x]$. On the one hand, $f(x)$ divides both $h(x)$ and $g(\alpha - cx)$ by the general factor theorem (Section 4.2). On the other hand, $f(x)$ has degree 1, since it has no multiple roots by the lemma and only one root is common to $h(x)$ and $g(\alpha - cx)$. It follows that $f(x) = x - \gamma$ and hence $\gamma \in F(\alpha)$. But then $\beta = \alpha - c\gamma$ is also in $F(\alpha)$, hence $F(\beta, \gamma) = F(\alpha)$ as required. □

Remarks

1. The theorem of the primitive element can be generalised far beyond number fields. As already indicated, the calculus in the proof of the lemma can be avoided by *defining* the derivative of $\Sigma a_i x^i$ to be $\Sigma i a_i x^{i-1}$ and proving the product rule algebraically. However, we also need the derivative to be nonzero, and for this the field F has to be one in which each element $ia_i \neq 0$ when $a_i \neq 0$. The term ia_i really means $a_i + a_i + \cdots + a_i$ (i times), so an equivalent statement is that each sum $1 + 1 + \cdots + 1 \neq 0$ in F. Such a field is said to be of *characteristic zero*. A field of characteristic zero obviously has infinitely many elements, which takes care of the other point needed for the proof, hence we can conclude that the theorem of the primitive element holds for any field of characteristic zero.

The function field $\mathbb{Q}(x_1, \ldots, x_n)$ is of characteristic zero, and it is of finite degree, n, over its subfield $\mathbb{Q}(a_0, \ldots, a_{n-1})$. Hence it follows from the generalised theorem that there is a primitive element $a_n \in \mathbb{Q}(x_1, \ldots, x_n)$ such that $\mathbb{Q}(x_1, \ldots, x_n) = \mathbb{Q}(a_0, \ldots, a_{n-1})(a_n)$. In fact, a_n can be any "totally asymmetric" function of $x_1, \ldots, x_n$, such as $x_1 + 2x_2 + 3x_3 + \cdots + nx_n$. This result is a tribute to the insight of Lagrange, Abel and Galois, since expressing $x_1, \ldots, x_n$ directly as rational functions of $a_0, \ldots, a_{n-1}$ and a chosen asymmetric a_n is extremely difficult, even for $n = 3$.

2. Counterexamples to the theorem occur in certain fields not of characteristic zero, but they are somewhat artificial. (See Dummit and Foote [1991], p.509, for one.)

3. In the case of finite fields, where both the lemma and the assumption of infinitely many elements break down, the theorem of the primitive element is

nevertheless true. In fact, we know that the nonzero elements of a finite field $F(\beta, \gamma)$ are *powers* of a single element α – a *primitive root* as it was called in Section 8.7*. This obviously implies $F(\beta, \gamma) = F(\alpha)$.

Exercises

9.1.1 Show that $\sqrt{2} + \sqrt{3}$ is a primitive element for $\mathbb{Q}(\sqrt{2}, \sqrt{3})$.

9.1.2 Show that if $\alpha_1, \ldots, \alpha_k$ are algebraic numbers, then there are integers $n_1, \ldots, n_k$ such that $\mathbb{Q}(\alpha_1, \ldots, \alpha_k) = \mathbb{Q}(n_1\alpha_1 + \cdots + n_k\alpha_k)$.

9.1.3 Prove $(fg)' = fg' + gf'$ algebraically, taking the derivative of $f(x) = \Sigma a_i x^i$ to be $f'(x) = \Sigma i a_i x^{i-1}$ by definition.

9.1.4 If $f(x) = (x - \alpha_1) \cdots (x - \alpha_n)$, where $\alpha_1, \ldots, \alpha_n \in \mathbb{C}$, find an expression for $f'(x)$ and hence show that

$$(\Delta(\alpha_1, \ldots, \alpha_n))^2 = \pm f'(\alpha_1) \cdots f'(\alpha_n),$$

where Δ^2 is the discriminant defined in Section 7.10 and the sign is $(-1)^{n(n-1)/2}$.

9.1.5 Deduce from Exercise 9.1.4 that the discriminant of $f(x) = x^n - 1$ is $\pm n^n$.

9.1.6 Use Exercise 9.1.5 and the fact that $\Delta \in \mathbb{Q}(\zeta_n)$ to deduce that $\mathbb{Q}(\zeta_n)$ contains the field $\mathbb{Q}(\sqrt{\pm n})$ when n is odd, in which case the sign is $(-1)^{(n-1)/2}$.

9.1.7 Show that $\sqrt{2} \in \mathbb{Q}(\zeta_8)$, and hence conclude that each quadratic field $\mathbb{Q}(\sqrt{n})$, where $n \in \mathbb{Z}$, is contained in some $\mathbb{Q}(\zeta_q)$ (a special case of the Kronecker-Weber theorem mentioned in Section 8.8).

9.2 Conjugate Fields and Splitting Fields

We know of cases where the Galois group $\mathrm{Gal}(E : F)$ fails to give any information about the structure of the field E over F. For example, $\mathrm{Gal}(\mathbb{Q}(\sqrt[3]{2}) : \mathbb{Q}) = \{1\}$ even though $\mathbb{Q}(\sqrt[3]{2})$ has degree 3 over $\mathbb{Q}$. In other cases, $\mathrm{Gal}(E : F)$ does seem to capture the structure of E over F. For example, $\mathrm{Gal}(\mathbb{Q}(\sqrt{2}) : \mathbb{Q})$ has two elements and the degree of $\mathbb{Q}(\sqrt{2})$ over $\mathbb{Q}$ is 2. We can explain and remove this discrepancy by looking beyond the automorphisms of E to the *isomorphisms* of E over F.

A field $E \subseteq \mathbb{C}$ may be isomorphic to various subfields of $\mathbb{C}$, and the various fields $\sigma(E) \subseteq \mathbb{C}$ onto which E is mapped by isomorphisms σ are called *conjugate fields* of E. As we shall see, they correspond to *conjugate elements* $\sigma(\alpha)$, in the sense of Section 6.2, where α is a primitive element for E over F. If all isomorphisms are onto E we call E *self-conjugate* or *normal*. As usual, we relativise these notions to a subfield F of E by considering the isomorphisms σ that fix F. Then the fields $\sigma(E)$ are called conjugate fields of E *over* F, and if they all equal E we call E *normal over* F.

For example, we get the isomorphisms σ of $\mathbb{Q}(\sqrt[3]{2})$ by setting $\sigma(\sqrt[3]{2})$ equal to $\sqrt[3]{2}$, $\zeta_3\sqrt[3]{2}$ or $\zeta_3^2\sqrt[3]{2}$. These are the only possibilities since σ must fix all rationals and therefore $(\sigma(\sqrt[3]{2}))^3 = \sigma(2) = 2$. Thus there are three isomorphisms of $\mathbb{Q}(\sqrt[3]{2})$, correctly reflecting the degree $(\mathbb{Q}(\sqrt[3]{2}):\mathbb{Q})$. The Galois group does not pick them up because they are onto three different conjugate fields, $\mathbb{Q}(\sqrt[3]{2})$, $\mathbb{Q}(\zeta_3\sqrt[3]{2})$ and $\mathbb{Q}(\zeta_3^2\sqrt[3]{2})$.

The isomorphisms of E over F always reflect $(E:F)$ correctly, as we see by generalising the above argument.

Lemma. *If $E, F \subseteq \mathbb{C}$ are fields with $(E:F)$ finite then the number of isomorphisms of E over F is $(E:F)$.*

Proof. Let $E = F(\alpha)$ by the theorem of the primitive element (Section 8.1) and let $p(x) \in F[x]$ be an irreducible polynomial satisfied by α. Then

$$\begin{aligned}(E:F) &= \text{degree}(p) \quad \text{by Section 5.4}\\ &= \text{number of distinct roots of } p(x)\end{aligned}$$

by Lemma 8.1 and the fundamental theorem of algebra (Section 3.8). On the other hand, each isomorphism σ of $E = F(\alpha)$ over F is given by the value of $\sigma(\alpha)$ (see Section 6.2), and it follows by applying σ to $p(\alpha) = 0$ that $p(\sigma(\alpha)) = 0$, since σ fixes the coefficients of $p(x)$. Thus $\sigma(\alpha)$ is one of the $(E:F)$ roots α_i of $p(x)$. Conversely, each α_i determines an isomorphism σ_i with $\sigma_i(\alpha) = \alpha_i$, namely the composite of the isomorphisms

$$F(\alpha) \to F[x]/p(x)F[x] \to F(\alpha_i)$$

we get from Section 6.2. Thus the number of isomorphisms is $(E:F)$. $\square$

The proof of the lemma also helps to explain why only isomorphisms reflect the degree of $\mathbb{Q}(\sqrt[3]{2})$, whereas automorphisms suffice for $\mathbb{Q}(\sqrt{2})$. Only one of the three isomorphisms of $\mathbb{Q}(\sqrt[3]{2})$ is onto $\mathbb{Q}(\sqrt[3]{2})$ because only the root $\sqrt[3]{2}$ of $p(x) = x^3 - 2$ is *in* $\mathbb{Q}(\sqrt[3]{2})$. The other two roots, $\zeta_3\sqrt[3]{2}$ and $\zeta_3^2\sqrt[3]{2}$, are outside $\mathbb{Q}(\sqrt[3]{2})$ and hence the corresponding isomorphisms are onto different conjugate fields. The irreducible polynomial $x^2 - 2$ satisfied by $\sqrt{2}$, on the other hand, has both its roots $\pm\sqrt{2}$ *in* $\mathbb{Q}(\sqrt{2})$, hence the corresponding isomorphisms are onto $\mathbb{Q}(\sqrt{2})$ and hence are automorphisms.

This explanation generalises beautifully to isomorphisms of E over F. We say that E is a *splitting field (of f) over F* if $E = F(\alpha_1, \ldots, \alpha_n)$ where $\alpha_1, \ldots, \alpha_n$ are the roots of some $f(x) \in F[x]$ (in which case $f(x)$ factorises completely, or "splits," into linear factors over E; this is the reason for the name). Then we have the following theorem.

Theorem. *If $E, F \subseteq \mathbb{C}$ are fields with $(E:F)$ finite then*

$$E \text{ is normal over } F \Leftrightarrow E \text{ is a splitting field over } F.$$

Proof. ($\Rightarrow$) Let $E = F(\alpha)$ and let $p(x) \in F[x]$ be an irreducible polynomial satisfied by α. It follows as in the lemma that each isomorphism σ_i of E over F is given by $\sigma_i(\alpha) = \alpha_i$ where α_i is a root of $p(x)$. Since E is normal over F, that is, each $\sigma_i(E) = E$, it follows that each root $\alpha_i = \sigma_i(\alpha)$ of $p(x)$ is in E, and hence $E = F(\alpha) = F(\alpha_1, \ldots, \alpha_n)$ is the splitting field of $p(x)$ over F.

($\Leftarrow$) Suppose $E = F(\alpha_1, \ldots, \alpha_n)$ is the splitting field of $p(x) \in F[x]$, so $\alpha_1, \ldots, \alpha_n$ are roots of $p(x)$. Applying an automorphism σ to each $p(\alpha_i) = 0$ as in the lemma we find that $\sigma(\alpha_i)$ is a root of $p(x)$ and hence $\sigma(\alpha_i) \in E$ since E is the splitting field of $p(x)$. Then, since any $\epsilon \in E$ is a rational function in the α_i with coefficients in F (Section 5.3), it follows that $\sigma(E) \subseteq E$, that is, $\sigma(E) = E$. □

Exercises

9.2.1 Why is $\mathbb{Q}(\zeta_3 \sqrt[3]{2}) \neq \mathbb{Q}(\zeta_3^2 \sqrt[3]{2})$?

9.2.2 Show that E is normal over $F \Leftrightarrow |\mathrm{Gal}(E : F)| = (E : F)$.

9.2.3 Show that a field E of finite degree over F is contained in a field E^* which is normal over F.

9.2.4 Use the isomorphisms of $\mathbb{Q}(\sqrt{2}, \sqrt{3})$ (Section 6.3)
(i) to show that $\mathbb{Q}(\sqrt{2}, \sqrt{3})$ is normal over $\mathbb{Q}$,
(ii) to find all four roots of the minimal polynomial for $\sqrt{2} + \sqrt{3}$.

9.2.5 Express $\mathbb{Q}(\sqrt[3]{2}, \zeta_3)$ as a splitting field.

9.2.6 Illustrate the ($\Leftarrow$) direction of the theorem above with the fields $E = \mathbb{Q}(x_1, \ldots, x_n)$ and $F = \mathbb{Q}(a_0, \ldots, a_{n-1})$, and compare with Section 6.5*.

9.3 Fixed Fields

The group of automorphisms of a field E fixing a subfield F is of course a crucial concept of Galois theory, the Galois group $\mathrm{Gal}(E : F)$. Just as crucial is the *fixed field* of a subgroup H of $\mathrm{Gal}(E : H)$:

$$\mathrm{Fix}(H) = \{\epsilon \in E : \sigma(\epsilon) = \epsilon \ \text{ for all } \sigma \in H\}.$$

$\mathrm{Fix}(H)$ is indeed a field because if σ fixes α, β then σ fixes $\alpha + \beta$, $\alpha - \beta$, $\alpha\beta$ and α/β. And Fix(H) is necessarily *between* E and F, that is, $E \supseteq \mathrm{Fix}(H) \supseteq F$, so we call it an *intermediate* field. Conversely, for any intermediate field B (B for "between") the group $\mathrm{Gal}(E : B)$ is necessarily a subgroup of $\mathrm{Gal}(E : F)$, since isomorphisms of E fixing B necessarily fix F. The same remark shows that a field E normal over F is also normal over B, a fact which is important in the proof of the theorem below. This theorem shows that the concepts of Galois group and fixed field are "inverse" to each other in the case of a field E normal over F.

First Fundamental Theorem of Galois Theory. *If $E, F \subseteq \mathbb{C}$ are fields with E normal over F then*
(i) $\mathrm{Fix}(\mathrm{Gal}(E : B)) = B$ *for each field* $B, E \supseteq B \supseteq F$,
(ii) $\mathrm{Gal}(E : \mathrm{Fix}(H)) = H$ *for each subgroup H of* $\mathrm{Gal}(E : F)$.

Proof. (i) It is immediate from the definition of fixed field that if $B' = \mathrm{Fix}(\mathrm{Gal}(E : B))$ then $B' \supseteq B$. Suppose that $\alpha \in B' - B$. Thus every automorphism of E fixing B also fixes α. On the other hand, since $\alpha \notin B$ the degree of α over B is ≥ 2, hence there are at least two isomorphisms τ of $B(\alpha)$ over F, sending α to the different roots of the $p(x) \in F[x]$ satisfied by α. For one of these, $\tau(\alpha) \neq \alpha$.

Now, viewing E as $B(\alpha)(\beta)$ by the theorem of the primitive element (Section 9.1), extend τ to an isomorphism σ of E (Section 6.3). Since τ fixes B so does σ, and since E is normal over F it is normal over B. This means σ is onto E and hence an automorphism, contradicting the assumption that every automorphism of E fixing F also fixes α.

(ii) Let $H = \{\sigma_1, \ldots, \sigma_r\}$, let $E = F(\alpha)$ and consider

$$f(x) = (x - \sigma_1(\alpha)) \cdots (x - \sigma_r(\alpha)).$$

For each $\sigma \in H$ the set $\{\sigma\sigma_1, \ldots, \sigma\sigma_r\} = \{\sigma_1, \ldots, \sigma_r\}$ (compare with Cayley's theorem, Section 7.2), hence we also have

$$f(x) = (x - \sigma\sigma_1(\alpha)) \cdots (x - \sigma\sigma_r(\alpha)).$$

This shows that the coefficients of $f(x)$ are fixed by each $\sigma \in H$, and hence $f(x) \in \mathrm{Fix}(H)[x]$.

It follows that $\sigma_1(\alpha), \ldots, \sigma_r(\alpha)$ include all the conjugates of α over $\mathrm{Fix}(H)$, hence $\sigma_1, \ldots, \sigma_r$ are all the isomorphisms of $E = \mathrm{Fix}(H)(\alpha)$ over $\mathrm{Fix}(H)$. Since they are in fact automorphisms we have

$$\mathrm{Gal}(E : \mathrm{Fix}(H)) = \{\sigma_1, \ldots, \sigma_r\} = H. \qquad \square$$

The fact that the maps $B \mapsto \mathrm{Gal}(E : B)$ and $H \mapsto \mathrm{Fix}(H)$ are inverses of each other means that, for normal extensions E of F, there is a one-to-one correspondence between intermediate fields B, $E \supseteq B \supseteq F$, and subgroups H of $\mathrm{Gal}(E : F)$. This correspondence is called the *Galois correspondence*. The first fundamental theorem is in fact usually expressed as the existence of the Galois correspondence.

Remark It follows from part (ii) of the theorem that any finite group H can be realised as $\mathrm{Gal}(E : F)$ for number fields $E \supseteq F$. First one uses Cayley's theorem to realise H as a subgroup of S_n, and hence as a subgroup of S_p for any $p > n$. Then one constructs an irreducible polynomial $f(x) \in \mathbb{Q}[x]$ of prime degree $p > n$ with exactly two nonreal roots, as in Section 8.5*. The argument used in Section 8.6* for $p = 5$ shows quite generally that the automorphism group of the root field E of $f(x)$ is S_p. Finally, if we take F to be the fixed field of H, viewed as a subgroup of this automorphism group, then we get $\mathrm{Gal}(E : F) = H$, as required.

The great open question of Galois theory is whether an arbitrary finite group H can be realised as $\mathrm{Gal}(E : \mathbb{Q})$. Some of the results obtained so far will be discussed in Section 9.9.

Exercises

9.3.1 Find the three intermediate fields between $\mathbb{Q}(\sqrt{2}, \sqrt{3})$ and $\mathbb{Q}$, and the corresponding subgroups of $\mathrm{Gal}(\mathbb{Q}(\sqrt{2}, \sqrt{3}) : \mathbb{Q})$.

9.3.2 Show that if E is of finite degree over F (not necessarily normal) then there are only finitely many intermediate fields.

9.3.3 Deduce from the proof of the theorem that $(E : B) = |\mathrm{Gal}(E : B)|$ for any field B between E and F, where E is normal over F.

9.3.4 Deduce from Exercise 9.2.6 and the Galois correspondence that any symmetric rational function of $x_1, \ldots, x_n$ is a rational function of the elementary symmetric functions $a_0, \ldots, a_{n-1}$.

9.3.5 Does it easily follow that a symmetric *polynomial* in $x_1, \ldots, x_n$ is a *polynomial* in $a_0, \ldots, a_{n-1}$? (Remember, this was the result of Exercise 6.5.1.)

9.4 Conjugate Intermediate Fields

It is no coincidence that the word "normal" applies to fields as well as groups. We are about to show that they denote matching concepts under the Galois correspondence. Actually we shall prove the more general result that conjugate intermediate fields correspond to conjugate subgroups, where subgroups H and H' of G are called *conjugate* when

$$H' = gHg^{-1} \text{ for some } g \in G.$$

Thus a normal subgroup (Section 7.8) is simply a self-conjugate subgroup, and the correspondence between normal intermediate fields and normal subgroups is established by the following theorem.

Second Fundamental Theorem of Galois Theory. *If $E, F \subseteq \mathbb{C}$ are fields with E normal over F then intermediate fields B, B' are conjugate over F $\Leftrightarrow$ $\mathrm{Gal}(E : B)$, $\mathrm{Gal}(E : B')$ are conjugate subgroups of $\mathrm{Gal}(E : F)$.*

Proof. ($\Rightarrow$) Suppose $B' = \tau(B)$, where τ is an isomorphism of B over F. Suppose that $E = F(\alpha)$ (by Section 9.1) and hence $E = B(\alpha)$. Then by Section 6.3 we can extend τ to an isomorphism σ of E over F, and since E is normal over F, σ is in fact an automorphism, that is, $\sigma \in \mathrm{Gal}(E : F)$. Thus $B' = \sigma(B)$ and I claim that $\mathrm{Gal}(E : \sigma(B)) = \sigma\mathrm{Gal}(E : B)\sigma^{-1}$. Indeed,

$$\begin{aligned}
\chi \in \mathrm{Gal}(E : \sigma(B)) &\Leftrightarrow \chi \in \mathrm{Gal}(E : F) \text{ and } \chi(\beta') = \beta' \text{ for all } \beta' \in \sigma(B) \\
&\Leftrightarrow \chi \in \mathrm{Gal}(E : F) \text{ and } \chi(\sigma(\beta)) = \sigma(\beta) \text{ for all } \beta \in B \\
&\Leftrightarrow \chi \in \mathrm{Gal}(E : F) \text{ and } \sigma^{-1}\chi\sigma(\beta) = \beta \text{ for all } \beta \in B \\
&\Leftrightarrow \sigma^{-1}\chi\sigma \in \mathrm{Gal}(E : B)
\end{aligned}$$

($\Leftarrow$ follows because an element of $\mathrm{Gal}(E : B)$ necessarily belongs to $\mathrm{Gal}(E : F)$)

$$\Leftrightarrow \chi \in \sigma\mathrm{Gal}(E : B)\sigma^{-1}.$$

($\Leftarrow$) Suppose $\mathrm{Gal}(E : B') = \sigma\mathrm{Gal}(E : B)\sigma^{-1}$ for some $\sigma \in \mathrm{Gal}(E : F)$. It follows from the equivalence just worked out that $\sigma\mathrm{Gal}(E : B)\sigma^{-1} = \mathrm{Gal}(E : \sigma(B))$, and hence $\mathrm{Gal}(E : B') = \mathrm{Gal}(E : \sigma(B))$. But then it follows from the Galois correspondence that $B' = \sigma(B)$, that is, B' is conjugate to B. □

Corollary. *If $E, B, F \subseteq \mathbb{C}$ are fields with E normal over F and $E \supseteq B \supseteq F$ then B is normal over F $\Leftrightarrow$ Gal$(E : B)$ is a normal subgroup of Gal$(E : F)$, in which case* $\text{Gal}(E : F)/\text{Gal}(E : B) \cong \text{Gal}(B : F)$.

Proof. The normality equivalence is immediate from the fact that normality is the same as self-conjugacy for both groups and fields. The group isomorphism $\text{Gal}(E : F)/\text{Gal}(E : B) \cong \text{Gal}(B : F)$ results from a natural homomorphism from $\text{Gal}(E : F)$ onto $\text{Gal}(B : F)$ with kernel $\text{Gal}(E : B)$ (compare with Section 7.9, Example 2).

The homomorphism is the *restriction to B* map $|_B$. For any $\sigma \in \text{Gal}(E : F)$ we define $\sigma|_B$ to be the function on domain B that agrees with σ there. It is routine to check that $\sigma|_B$ is an isomorphism of B over F, and since B is assumed normal over F it follows that $\sigma|_B$ is onto B and hence $\sigma\ |_B \in \text{Gal}(B : F)$. Conversely, any $\tau \in \text{Gal}(B : F)$ extends to $\sigma \in \text{Gal}(E : F)$ by the argument in the $(\Rightarrow)$ direction of the theorem, hence $|_B$ is from $\text{Gal}(E : F)$ *onto* $\text{Gal}(B : F)$. And it is a homomorphism because if $\sigma, \sigma' \in \text{Gal}(E : F)$ we have

$$\sigma\sigma'|_B(\beta) = \sigma|_B\sigma'|_B(\beta) \quad \text{since } \sigma|_B(\beta) \in B \text{ for all } \beta \in B.$$

Finally,

$$\ker(|_B) = \{\sigma : \sigma \in \text{Gal}(E : F), \sigma \text{ fixes } B\} = \text{Gal}(E : B)$$

as required. □

What we have called the first and second fundamental theorems are often combined into a single "fundamental theorem of Galois theory." There is also an enhancement, which we shall not need, equating the relative dimension of intermediate fields with the relative size of the corresponding Galois groups. To be precise, if $B_1 \supseteq B_2$ are fields between F and its normal extension E, and if

$$H_1 = \text{Gal}(E : B_1),\ H_2 = \text{Gal}(E : B_2),$$

then

$$(B_1 : B_2) = [H_2 : H_1].$$

This is easily deduced from Exercise 9.3.3, the Dedekind product theorem, and the corresponding index product theorem (Section 7.7).

Exercises

9.4.1 Find the three conjugate non-normal subgroups of $\text{Gal}(\mathbb{Q}(\sqrt[3]{2}, \zeta_3) : \mathbb{Q}) = S_3$ (compare with Section 7.4), and the corresponding conjugate fields.

9.4.2 Find a field between $\mathbb{Q}(\sqrt[3]{2}, \zeta_3)$ and $\mathbb{Q}$ which is normal over $\mathbb{Q}$.

9.4.3 Prove that $(B_1 : B_2) = [H_2 : H_1]$, where B_1, B_2, H_1, H_2 are as described above.

9.4.4 Use the enhancement of the fundamental theorem, and the fact that $\text{Gal}(\mathbb{Q}(\zeta_p) : \mathbb{Q}) \cong C_{p-1}$ for an odd prime p (Why?), to show that $\mathbb{Q}(\zeta_p)$ contains a unique field E with $(E : \mathbb{Q}) = 2$.

9.4.5 Deduce from Exercises 9.4.4 and 9.1.6 that the quadratic field $E \subset \mathbb{Q}(\zeta_p)$ is $\mathbb{Q}(\sqrt{\pm p})$.

9.4.6 Use Exercise 1.5.2 to show in particular that $\mathbb{Q}(\sqrt{5}) \subset \mathbb{Q}(\zeta_5)$.

9.5 Normal Extensions with Solvable Galois Group

We now have enough Galois theory to be able to prove a converse to the necessary condition for solvability of equations found in Chapter 8. To make this a true converse we shall first revise the necessary condition for solvability so that it becomes a condition on normal extensions.

If $f(x) \in \mathbb{Q}[x]$ is solvable by radicals then the argument of Section 8.3 shows that the field $\mathbb{Q}(\alpha_1, \ldots, \alpha_n)$, where $\alpha_1, \ldots, \alpha_n \in \mathbb{C}$ are the roots of $f(x)$, has solvable Galois group over $\mathbb{Q}$. Since $\mathbb{Q}(\alpha_1, \ldots, \alpha_n)$ is the splitting field for $f(x)$ we can now add, by Section 9.3, that $\mathbb{Q}(\alpha_1, \ldots, \alpha_n)$ is normal over $\mathbb{Q}$. Hence *if $f(x) \in \mathbb{Q}[x]$ and $f(x) = 0$ is solvable by radicals then the roots of $f(x)$ lie in a normal extension of $\mathbb{Q}$ with solvable Galois group.* To prove the converse, that if the roots lie in a normal extension of $\mathbb{Q}$ with solvable Galois group then $f(x) = 0$ is solvable by radicals, it therefore suffices to show that *any normal extension of $\mathbb{Q}$ with solvable Galois group is contained in a radical extension of $\mathbb{Q}$.*

In the present section we shall use Corollary 9.4 to reduce this assertion to the special case of an extension with cyclic Galois group.

Theorem. *If E, F are number fields and E is normal over F with Gal$(E : F)$ solvable, then there are fields $F = F_0 \subseteq F_1 \subseteq \ldots \subseteq F_k = E$ with each F_i normal over F_{i-1} and Gal$(F_i : F_{i-1})$ cyclic of prime order.*

Proof. Solvability of $\text{Gal}(E : F)$ means that there are groups $\text{Gal}(E : F) = G_0 \supseteq G_1 \ldots \supseteq G_k = \{1\}$ with each G_i normal in G_{i-1} and G_{i-1}/G_i cyclic of prime order (Section 8.8). Let $F_i = \text{Fix}(G_i)$. Then

$$F_0 \subseteq F_1 \subseteq \ldots \subseteq F_k = E$$

where $F = F_0$ by the Galois correspondence since E is normal over F. E is also normal over each F_i, and F_{i-1}. The Galois correspondence says $\text{Gal}(E : F_i) = G_i$, which is therefore a normal subgroup of $\text{Gal}(E : F_{i-1}) = G_{i-1}$. It follows, by Corollary 9.4, that F_i is normal over F_{i-1} and hence, by Corollary 9.4 again,

$$\text{Gal}(F_i : F_{i-1}) \cong \text{Gal}(E : F_{i-1})/\text{Gal}(E : F_i) = G_{i-1}/G_i,$$

which is cyclic of prime order by hypothesis. □

9.6 Cyclic Extensions

Having broken down our normal solvable extension E of F into steps $F = F_0 \subseteq F_1 \subseteq \ldots \subseteq F_k = E$ such that each F_i is normal over F_{i-1} and $\mathrm{Gal}(F_i : F_{i-1})$ is cyclic of prime order, it remains to show that each such step is "small enough" to be described by adjunction of a pth root. The situation is the mirror image of the one described in Section 8.3, where a radical extension was broken down into adjunctions of pth roots, and these steps proved "small enough" to be described group-theoretically.

Unfortunately, we again have trouble with roots of unity. We are only able to prove that F_i is a radical extension of F_{i-1} when $\zeta_p \in F_{i-1}$, where p is the order of $\mathrm{Gal}(F_i : F_{i-1})$. We shall carry out the proof of this special case here, then use it in Section 9.7 to show that any normal solvable extension is *contained in* a radical extension, which meets our goal of Section 9.5. The present situation is definitely more difficult than that of Section 8.3. There we knew what the group was and had only to prove it abelian. Here we have no idea what the pth root is, and it is hard to believe it can be conjured out of $C_p = \mathrm{Gal}(F_i : F_{i-1})$. An ingenious way of doing this, with the help of ζ_p, was discovered by Lagrange [1771].

Theorem. *If F_i is normal over F_{i-1}, $\mathrm{Gal}(F_i : F_{i-1}) \cong C_p$ for p prime, and $\zeta_p \in F_{i-1}$, then $F_i = F_{i-1}(\sqrt[p]{\beta})$ for some $\beta \in F_{i-1}$.*

Proof. Since $\mathrm{Gal}(F_i : F_{i-1})$ has p elements, and all isomorphisms of F_i over F_{i-1} are onto F_i by normality, it follows by Lemma 9.3 that $(F_i : F_{i-1}) = p$. Since p is prime it follows from the Dedekind product theorem that there is no proper intermediate field, and hence $F_i = F_{i-1}(\alpha)$ for *any* $\alpha \in F_i - F_{i-1}$. Thus it suffices to find some $\sqrt[p]{\beta} \in F_i - F_{i-1}$ with $\beta \in F_{i-1}$.

Choose any $\alpha \in F_i - F_{i-1}$, so $F_i = F_{i-1}(\alpha)$, and let $\mathrm{Gal}(F_i : F_{i-1}) = \{1, \sigma, \sigma^2, \ldots, \sigma^{p-1}\}$. Consider the numbers

$$\alpha_j = \alpha + \zeta_p^j \sigma(\alpha) + \cdots + \zeta_p^{(p-1)j} \sigma^{p-1}(\alpha) \quad \text{for} \quad 0 \leq j \leq p-1. \qquad (*)$$

Since $\zeta_p \in F_{i-1}$ is fixed by σ we have

$$\begin{aligned} \sigma(\alpha_j) &= \sigma(\alpha) + \zeta_p^j \sigma^2(\alpha) + \cdots + \zeta_p^{(p-1)j} \alpha \\ &= \zeta_p^{-j} \alpha_j, \end{aligned}$$

and therefore

$$\sigma(\alpha_j^p) = (\zeta_p^{-j} \alpha_j)^p = \zeta_p^{-jp} \alpha_j^p = \alpha_j^p.$$

Thus α_j^p belongs to the fixed field of σ, which is F_{i-1} by the normality of F_i over F_{i-1} and the Galois correspondence.

This suggests we take $\beta = \alpha_j^p$, however, this is no use if $\alpha_j \in F_{i-1}$, which is why we have left the value of j open so far. We can show that *some* $\alpha_j \notin F_{i-1}$ by the following trick.

Notice that when $k \neq 0$ the set $\{1, \zeta_p^k, \ldots, \zeta_p^{(p-1)k}\}$ of coefficients in the kth "column" of (*) equals $\{1, \zeta_p, \ldots, \zeta_p^{p-1}\}$ since p is prime, and hence sums to 0 by the cyclotomic equation (Section 3.7). Thus we have

$$\sum_{j=0}^{p-1} \alpha_j = p\alpha \notin F_{i-1}.$$

It follows that some $\alpha_j \notin F_{i-1}$, so we take $\beta = \alpha_j^p$ for this particular j. □

9.7 Construction of the Radical Extension

At last we are in a position to construct a radical extension containing a given normal solvable extension. In Section 9.5 we broke down the given extension E of F into steps

$$F = F_0 \subseteq F_1 \subseteq \ldots \subseteq F_k = E$$

with each F_i normal over F_{i-1} and $\mathrm{Gal}(F_i : F_{i-1})$ cyclic of prime order, p_i, say. In Section 9.6 we showed how to find a $\beta_i \in F_{i-1}$ such that $F_i = F_{i-1}(\sqrt[p_i]{\beta_i})$ in the special case where $\zeta_{p_i} \in F_{i-1}$.

We now construct a radical extension of F which *contains* E by retracing the steps $F_0, F_1, \ldots, F_k$, but adjoining numbers ζ_{p_i} where needed. The key to the successful adjunction of numbers ζ_{p_i} is the following lemma.

Lemma. *If F_i is normal over F_{i-1} with $(F_i : F_{i-1}) = p_i$ then $F_i(\zeta_{p_i})$ is normal over $F_{i-1}(\zeta_{p_i})$ with $(F_i(\zeta_{p_i}) : F_{i-1}(\zeta_{p_i})) = p_i$.*

Proof. F_i normal over F_{i-1}

$$\begin{aligned}
&\Rightarrow F_i \text{ is the splitting field of some } f(x) \text{ over } F_{i-1} \text{ by Section 9.2}\\
&\Rightarrow F_i(\zeta_{p_i}) \text{ is the splitting field of } (x^{p_i}-1)f(x) \text{ over } F_{i-1}\\
&\Rightarrow F_i(\zeta_{p_i}) \text{ normal over } F_{i-1} \text{ by Section 9.2}\\
&\Rightarrow F_i(\zeta_{p_i}) \text{ normal over } F_{i-1}(\zeta_{p_i}) \text{ by definition of normality.}
\end{aligned}$$

Now if $\epsilon_1, \ldots, \epsilon_m$ is a basis for F_i over F_{i-1} it follows from the Dedekind product theorem that the elements $\epsilon_j\zeta_{p_i}^k$ span $F_i(\zeta_{p_i})$ over F_{i-1}, and hence over $F_{i-1}(\zeta_{p_i})$. But the $\zeta_{p_i}^k \in F_{i-1}(\zeta_{p_i})$, so in fact $\epsilon_1, \ldots, \epsilon_m$ span $F_i(\zeta_{p_i})$ over $F_{i-1}(\zeta_{p_i})$, and therefore

$$(F_i(\zeta_{p_i}) : F_{i-1}(\zeta_{p_i})) \leq (F_i : F_{i-1}) = p_i$$

On the other hand, since $F_i(\zeta_{p_i}) \supseteq F_{i-1}(\zeta_{p_i}) \supseteq F_{i-1}$ and $F_i(\zeta_{p_i}) \supseteq F_i \supseteq F_{i-1}$ the Dedekind product theorem gives

$$(F_i(\zeta_{p_i}) : F_{i-1}) = (F_i(\zeta_{p_i}) : F_{i-1}(\zeta_{p_i}))(F_{i-1}(\zeta_{p_i}) : F_{i-1})$$

and also

$$(F_i(\zeta_{p_i}) : F_{i-1}) = (F_i(\zeta_{p_i}) : F_i)(F_i : F_{i-1}) = (F_i(\zeta_{p_i}) : F_i)p_i.$$

Then since $(F_{i-1}(\zeta_p) : F_{i-1}) \leq \mathrm{degree}(\zeta_{p_i}) = p_i - 1$ it follows that p_i divides $(F_i(\zeta_{p_i}) : F_{i-1}(\zeta_{p_i}))$, and therefore $(F_i(\zeta_{p_i}) : F_{i-1}(\zeta_{p_i})) = p_i$. □

Theorem. *If E is a normal solvable extension of F then E is contained in a radical extension of F.*

Proof. Let $F = F_0 \subseteq F_1 \subseteq \ldots \subseteq F_k = E$ be the fields given by Section 9.5, where F_i is normal over F_{i-1} and $\mathrm{Gal}(F_i : F_{i-1})$ is cyclic of prime order p_i. It follows that $(F_i : F_{i-1}) = p_i$ by Section 9.2, since the automorphisms of a normal extension are all its isomorphisms. Then it follows by the lemma that $F_i(\zeta_{p_i})$ is normal over $F_{i-1}(\zeta_{p_i})$ and $(F_i(\zeta_{p_i}) : F_{i-1}(\zeta_{p_i})) = p_i$.

It follows, in turn, that the isomorphisms of $F_i(\zeta_{p_i})$ over $F_{i-1}(\zeta_{p_i})$ are all automorphisms, and $|\mathrm{Gal}(F_i(\zeta_{p_i}) : F_{i-1}(\zeta_{p_i}))| = p_i$ by Section 9.2 again. Since p_i is prime this means $\mathrm{Gal}(F_i(\zeta_{p_i}) : F_{i-1}(\zeta_{p_i})) = C_{p_i}$, because C_{p_i} is the only group with p_i elements (a nontrivial cyclic subgroup cannot have less than p_i elements by Lagrange's theorem, Section 7.7). Then Section 9.5 gives

$$F_i(\zeta_{p_i}) = F_{i-1}(\zeta_{p_i}, \sqrt[p_i]{\beta_i}) \quad \text{for some} \quad \beta_i \in F_{i-1}(\zeta_{p_i}).$$

This shows that by extending $F = F_0$ by the successive radicals $\zeta_{p_1}, \sqrt[p_1]{\beta_1}, \zeta_{p_2}, \sqrt[p_2]{\beta_2}, \ldots, \zeta_{p_k}, \sqrt[p_k]{\beta_k}$ we obtain a radical extension containing $F_k = E$, as required. □

Corollary. *If $f(x) \in \mathbb{Q}[x]$ has a splitting field E over $\mathbb{Q}$, and $\mathrm{Gal}(E : \mathbb{Q})$ is solvable, then $f(x) = 0$ is solvable by radicals.*

Proof. Since any splitting field is normal (Section 9.2), it follows from the theorem that E is contained in a radical extension of $\mathbb{Q}$. In particular, the roots of $f(x) = 0$ are in this radical extension and hence are expressible in terms of rational numbers and radicals. □

9.8 Construction of Regular p-gons

In Section 5.6 we showed that, when p is prime, the regular p-gon is constructible only if $p-1$ is a power of 2. The proof used the fact that construction of the p-gon is equivalent to construction of ζ_p, and that $(\mathbb{Q}(\zeta_p) : \mathbb{Q}) = p - 1 = 2^m$ when p is constructible. Since $\mathbb{Q}(\zeta_p)$ is the splitting field of $x^p - 1$ over $\mathbb{Q}$, hence normal, we can now use properties of normal extensions to prove a converse theorem. In fact we can prove the following result.

Theorem. *If E is a normal extension of F with $(E : F) = 2^m$ and $\mathrm{Gal}(E : F)$ abelian then E is a radical extension of F by square roots only.*

Proof. We shall prove this for any number fields E and F by induction on m. The theorem is trivial for $m = 0$ and easy for $m = 1$. In the latter case $E = F(\alpha)$ by Section 9.1 and α satisfies a quadratic equation since $(E : F) = 2$, hence α is expressible by a single square root. We therefore suppose $m > 1$ and that the theorem is true for all $k < m$, and suppose we are given a normal E with $(E : F) = 2^m$ and $\mathrm{Gal}(E : F)$ abelian.

There are 2^m isomorphisms of E over F by Lemma 9.2 and since E is normal these isomorphisms are all onto E and hence $|\mathrm{Gal}(E : F)| = 2^m$. We now find a proper nontrivial subgroup H of $\mathrm{Gal}(E:F)$. If $\mathrm{Gal}(E:F)$ is cyclic, say $\{\mathbf{1},\sigma,\sigma^2,\ldots,\sigma^{2^m-1}\}$, let $H=\{\mathbf{1},\sigma^2,\sigma^4,\ldots,\sigma^{2^m-2}\}$. If $\mathrm{Gal}(E:F)$ is not cyclic let σ be a nonidentity element and let H be the cyclic subgroup consisting of the powers of σ.

Since E is normal over F, the field $\mathrm{Fix}(H)$ is strictly between E and F by the Galois correspondence (Section 9.3). This gives us extensions with

$$(E:\mathrm{Fix}(H))=2^k<2^m$$
$$(\mathrm{Fix}(H):F)=2^l<2^m$$

since

$$(E:\mathrm{Fix}(H))(\mathrm{Fix}(H):F)=(E:F)=2^m$$

by the Dedekind product theorem. Both of these extensions are normal. E is normal over $\mathrm{Fix}(H)$ because E is normal over F. $\mathrm{Fix}(H)$ is normal over F because it corresponds to the subgroup H of $\mathrm{Gal}(E:F)$, which is a normal subgroup because $\mathrm{Gal}(E:F)$ is abelian. Finally, $\mathrm{Gal}(E:\mathrm{Fix}(H))$ is abelian because it is a subgroup of $\mathrm{Gal}(E:F)$, and $\mathrm{Gal}(\mathrm{Fix}(H):F)$ is abelian because it is a quotient group of $\mathrm{Gal}(E:F)$ by Section 9.4.

Thus it follows by induction that $\mathrm{Fix}(H)$ is a radical extension of F by square roots only, and that E is a radical extension of $\mathrm{Fix}(H)$, and hence of F, by square roots only. This completes the induction. □

Corollary. *ζ_p is constructible if p is prime and $p=2^m+1$*

Proof. In this case $E=\mathbb{Q}(\zeta_p)$ has degree 2^m by Section 5.1, and its Galois group over $F=\mathbb{Q}$ is abelian by Section 8.3. Thus it follows from the theorem that ζ_p is expressible in terms of rational numbers and square roots. The square roots, however, may be applied to negative numbers. To show that ζ_p is constructible, that is, of the form $\zeta_p=\alpha+i\beta$ where α,β are in the *real* quadratic closure of $\mathbb{Q}$ (Section 1.2), we have to investigate the square roots of complex numbers.

Suppose

$$\sqrt{\gamma+i\delta}=\alpha+i\beta.$$

Then

$$\gamma+i\delta=(\alpha+i\beta)^2=\alpha^2-\beta^2+2i\alpha\beta,$$

and hence

$$\alpha^2-\beta^2=\gamma,\quad 2\alpha\beta=\delta.$$

Substituting $\beta=\delta/2\alpha$ in the first of these gives

$$4(\alpha^2)^2-4\gamma\alpha^2-\delta^2=0$$

which has the solution

$$\alpha=\sqrt{\frac{\gamma+\sqrt{\gamma^2+\delta^2}}{2}}$$

Since $\sqrt{\gamma^2+\delta^2}\geq|\gamma|$, this expression for α involves square roots only of positive reals, and hence so does the corresponding expression $\beta=\delta/2\alpha$. Induction on the number of square roots in $\zeta_p = \alpha + i\beta$ then shows that α, β are in the real quadratic closure of $\mathbb{Q}$. Thus ζ_p, and hence the regular p-gon, is constructible when $p = 2^m + 1$. □

Exercise

9.8.1 Give another proof that if $\alpha + i\beta$ is constructible so is $\sqrt{\alpha + i\beta}$, using the polar representation $\alpha + i\beta = re^{i\theta}$.

9.9* Division of Arbitrary Angles

Recall from Section 5.6 how we showed that trisection of arbitrary angles is not constructible. We took the *constructible* angle $\pi/3$ and showed that one-third of it, $\pi/9$, is not constructible. Now that we know $2\pi/p$ is constructible for any Fermat prime p, and that $2\pi/n$ is *not* constructible unless $n = 2^m p_1 \cdots p_k$ for distinct Fermat primes $p_1, \ldots, p_k$ (Section 5.7*), we are in a position to generalise the argument against trisection to the following:

Theorem. *Constructible n-section of arbitrary angles exists $\Leftrightarrow$ n is a power of 2.*

Proof. ($\Leftarrow$) If $n = 2^m$ then n-section reduces to a series of bisections, and bisection of an arbitrary angle is done by the construction shown in Figure 9.9.1.

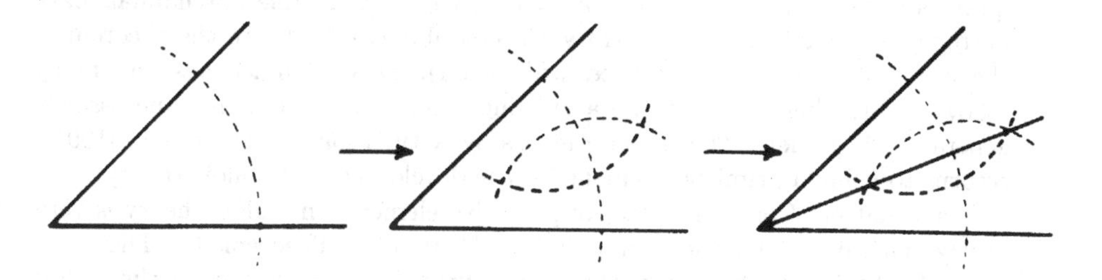

Fig. 9.9.1. Bisection of an angle

($\Rightarrow$) Conversely, if n is not a power of 2 then $n = pq$ for some odd prime p. It suffices to show that p-section of arbitrary angles is not constructible, because construction of an n-section implies construction of a p-section (by laying q copies of the n-sected angle side by side).

If p is not a Fermat prime then an angle which cannot be p-sected is 2π, since 2π is constructible but $2\pi/p$ is not (by Theorem 5.7*). If p is a Fermat prime then an angle which cannot be p-sected is $2\pi/p$, since $2\pi/p$ is constructible (by Theorem 9.8) but $2\pi/p^2$ is not (again by Theorem 5.7*). □

9.10 Discussion

The theorem of the primitive element, which is so important for the approach to Galois theory in this chapter, was also important for Galois. His version of it is the first substantial result in his memoir [1831]. It states that the root field $\mathbb{Q}(\alpha_1, \ldots, \alpha_n)$ of an irreducible $p(x) \in \mathbb{Q}[x]$ can be expressed in the form $\mathbb{Q}(n_1\alpha_1 + \cdots + n_k\alpha_k)$ where $n_1, \ldots, n_k$ are suitably chosen integers. One of his referees, Poisson, judged Galois' proof to be insufficient, but noted that it could be saved with the help of Lagrange [1771], no.100 (see Edwards [1984], p.103). Lagrange does indeed have a theorem which easily implies the existence of a primitive element (see Rotman [1990], p.99). Thus the theorem of the primitive element has a pedigree in Galois theory as long as that of the symmetric group. Even when Dedekind began to modernise Galois theory in his work [1893], the theorem of the primitive element retained a leading role (Theorem VI of [1893], §165). This was possible because Dedekind's concern, like ours in the present chapter, was with number fields.

Dedekind's successors, Emmy Noether and Emil Artin, were not content to restrict Galois theory to such fields, and saw the theorem of the primitive element as an obstacle to greater generality. According to Weyl [1935], Emmy Noether "disdained to employ a primitive element in the development of Galois theory." Artin [1942] found that by extending Dedekind's ideas from linear algebra it became possible to prove a "fundamental theorem of Galois theory" for general field extensions without use of a primitive element. His approach requires a division into the "separable case" (where irreducible polynomials have no repeated roots) and the contrary "inseparable case" (where the theorem of the primitive element is not necessarily valid). This approach has had many followers, but there seems to be a growing recognition that it is unnecessarily general for beginners. The recent *Algebra* by Artin's son Michael (Artin [1991]) returns to using a primitive element for the development of Galois theory.

As a matter of fact, the place of primitive elements in Galois theory is very neatly marked out by a theorem of Steinitz [1910] (§14, Theorem III). The inspiration for Steinitz's theorem is the rather surprising consequence of the Galois correspondence, for number fields, that a finite degree extension E of F has only finitely many intermediate fields. As suggested by Exercises 9.2.3 and 9.3.2, this is because E lies in a normal extension E^* of F, also of finite degree. It follows that $|\mathrm{Gal}(E^* : F)| = (E^* : F)$ is finite and, by normality, fields between E and F correspond to some of the finitely many subgroups of $\mathrm{Gal}(E^* : F)$. To construct the finite degree normal extension E^* one of course expresses E as $F(\alpha)$, and adjoins the finitely many images of α under isomorphisms of E. Use of a primitive element is unavoidable here, because it is in fact *equivalent* to the property being proved. Steinitz's theorem is that a finite degree extension E of F has finitely many intermediate fields if and only if $E = F(\alpha)$ for some $\alpha \in E$. (For an accessible proof see Rotman [1990], p.50.)

The problem of realising a given finite group as $\mathrm{Gal}(E : \mathbb{Q})$ was apparently posed by Dedekind, though I have been unable to find the statement of it in his works. It is relatively easy to prove that any finite abelian group can be realised in this way (see for example Dummit and Foote [1991], p.514). Perhaps the most impressive result so far is that of Shafarevich [1954], which says that every solvable group is realisable. This means that every group that *could* occur as the Galois group of a solvable equation actually does occur. It also means that the remaining "problem groups" are the nonsolvable groups, and particularly the simple groups. Now that we have a complete list of finite simple groups (assuming the 10,000+ page proof is correct!) there is keen interest in realising particular simple groups as Galois groups over $\mathbb{Q}$. For a report on recent progress, see Serre [1992].

The expressions $\alpha+\zeta_p^j\sigma(\alpha)+\cdots+\zeta_p^{(p-1)j}\sigma^{p-1}(\alpha)$ used in Section 9.6 are called *Lagrange resolvents* because expressions of this type were used, in a somewhat similar way, by Lagrange [1771]. In nos. 6 to 8 of this work he used them to give a symmetric solution of the general cubic equation. Here is the gist of his method. Suppose the roots of the cubic $x^3+a_2x^2+a_1x+a_0$ are x_1, x_2, x_3 and consider the resolvent

$$t_1 = x_1 + \zeta_2 x_2 + \zeta_2^2 x_3.$$

Let $t_2, \ldots, t_6$ be the terms obtained from t_1 by permuting x_1, x_2, x_3 nontrivially. Then the coefficients of

$$f(x) = (x-t_1)\cdots(x-t_6) \qquad (*)$$

are symmetric in x_1, x_2, x_3 and hence polynomials in a_0, a_1, a_2 by the fundamental theorem of symmetric functions (Section 6.5* and Exercise 9.3.4). Moreover, it turns out that $f(x)$ is quadratic in x^3, and hence its roots are expressible as radicals in a_0, a_1, a_2. Finally, Lagrange recovers the original roots x_1, x_2, x_3 as linear combinations of the t_is.

Expressions similar to Lagrange resolvents (but involving just roots of unity) were used by Gauss [1801] in his proof that the regular p-gon is constructible when p is a Fermat prime. As we know from Section 5.7*, this determines the constructible n-gons – namely, as those for which n is the product of a power of 2 by distinct Fermat primes. Unfortunately, we do not know whether there are any Fermat primes beyond 3, 5, 17, 257, 65537. Thus we have to admit that this classical geometric problem is not completely solved. We know something the Greeks didn't know – that the problem is really one about prime numbers, and that it can be solved in cases they couldn't solve – but we still don't know much about Fermat primes.

Deciding whether 65537 is the last Fermat prime may well tax the best mathematicians of the future. In the meantime, there is a smaller problem about 65537 which has also not been solved: express ζ_{65537} in terms of square roots! Expressions for ζ_{17} and ζ_{257} were worked out early in the 19th century, and a Professor Hermes of Göttingen is said to have spent 10 years in an unsuccessful attempt to compute ζ_{65537}. Apparently the computations became too big for him. Surely today, with computer algebra systems, this should no longer be a problem. If 65537 is indeed the last Fermat prime, whoever computes ζ_{65537} should earn at least a footnote in the history of mathematics.

References

N.H. Abel
[1826] Démonstration de l'impossibilitê de la résolution algébrique des équations générales qui passent le quatrieme degré. *J. Reine Angew. Math.* **1**, 65-84, *Œuvres Complètes* 1, 66-87. An English translation of a short version of this paper is in: *A Source Book in Mathematics*, vol.1 (Ed. D.E. Smith), Dover, New York, 1959.

N.H. Abel
[1827] Recherches sur les fonctions elliptiques. *J. Reine Angew. Math.* **2**, 101-181, 3, 160-190 *Œuvres Complètes* 1, 263-388.

N.H. Abel
[1829] Mémoire sur une classe particulière d'équations résolubles algébriquement. *J. Reine Angew. Math.* **4**, 131-156. *Œuvres Complètes* 1, 478-507.

R. Argand
[1806] *Essai sur une manière de représenter les quantités imaginaires dans les constructions géométriques.* Paris.

E. Artin
[1942] *Galois Theory.* University of Notre Dame, Notre Dame.

M. Artin
[1991] *Algebra.* Prentice-Hall, Englewood Cliffs, New Jersey.

C. Bachet
[1621] *Diophanti Alexandrini Arithmeticorum libri sex.* H. Drovart, Paris.

J. Bernoulli
[1694] Curvatura laminae elastica. *Acta Eruditorum* **13**, 262-276.

L. Bieberbach
[1952] *Theorie der Geometrischen Konstruktionen.* Birkhäuser, Basel.

B. Bolzano
[1817] Rein analytischer Beweis des Lehrsatzes dass zwischen je zwey Werthen, die ein entgegengesetzes Resultat gewähren, wenigstens eine reelle Wurzel der Gleichung liege. *Ostwald's Klassiker*, vol. 153, Engelmann, Leipzig, 1905.

R. Bombelli
[1572] *L'Algebra.* Feltrinelli Editore, Milan, 1966.

E. Brieskorn & H. Knörrer
[1986] *Plane Algebraic Curves.* Birkhäuser, Basel.

G. Cantor
[1874] Über eine Eigenschaft des Inbegriffes aller reellen algebraischen Zahlen. *J. Reine Angew. Math.* **77**, 258-262, *Gesammelte Abhandlungen*, 115-118.

G. Cantor
[1891] Über eine elementare Frage der Mannigfaltigkeitslehre. *Jahrb. Deutsch. Math. Ver.* **1**, 75-78, *Gesammelte Abhandlungen*, 278-280.

G. Cardano
[1545] *Ars Magna*. English translation: *The Great Art*. MIT Press, Cambridge, Mass., 1968.

A. Cauchy
[1815] Mémoire sur le nombre des valeurs qu'une fonction peut acquerir, lorsqu'on y permute de toutes les manières possibles les quantités qu'elle renferme. *J. Éc. Polytech.* **18**, 10, 1-28, *Œuvres*, ser. 2, 1, 62-90.

A. Cauchy
[1821] Cours d'analyse de l'école royale polytechnique. *Œuvres*, ser. 1, 3.

A. Cauchy
[1847] Mémoire sur la théorie des équivalences algébriques, substituée à la théorie des imaginaires. *Exercises d'analyse et de physique mathématique,* Tome 4, 87-110.

A. Cayley
[1854] On the theory of groups, as depending on the symbolic equation $\theta^n = 1$. *Phil. Mag.* **7**, 40-47, *Collected Mathematical Papers* 2, 123-130.

N.G. Čebotarev
[1934] Über quadrierbare Kreisbogen Zweiecke I. *Math. Zeit.* **39**, 161-175.

T. Clausen
[1840] Vier neue mondförmige Flächen, deren Inhalt quadrierbar ist. *J. Reine Angew. Math.* **21**, 375-376.

R. Cotes
[1722] *Harmonia Mensurarum*. Cambridge.

H.S.M. Coxeter
[1961] *Introduction to Geometry*. Wiley, New York.

J.L. d'Alembert
[1746] Recherches sur le calcul integral. *Hist. Acad. Sci. Berlin*, **2**, 182-224.

A. de Moivre
[1707] Equationum quarundum potestatis tertiae, quintae, septimae, nonae & superiorum, ad infinitum usque pergendo, in terminus finitis, ad instar regularum pro cubicus que vocantur Cardani, resolutio analytica. *Phil. Trans.* **25**, 2368-2371.

R. Dedekind
[1857] Abriss einer Theorie der höheren Kongruenzen in bezug auf einen reellen Primzahl-Modulus. *J. Reine Angew. Math.* **54**, 1-26, *Werke* I, 40-67.

R. Dedekind
[1857′] Beweis für die Irreduktibilität der Kreisteilungs-Gleichung. *J. Reine Angew. Math.* **54**, 27-30. *Werke* I, 68-71.

R. Dedekind
[1871] Supplement X to Dirichlet's *Vorlesungen über Zahlentheorie*, 2nd Ed. Vieweg, Braunschweig.

R. Dedekind
[1872] *Stetigkeit und die Irrationalzahlen.* English translation in *Essays on the Theory of Numbers.* Open Court, Chicago, 1901.

R. Dedekind
[1888] *Was sind und was sollen die Zahlen?* English translation in *Essays on the Theory of Numbers.* Open Court, Chicago, 1901.

R. Dedekind
[1893] Supplement XI to Dirichlet's *Vorlesungen über Zahlentheorie*, 4th Ed. Vieweg, Brauschweig.

R. Descartes
[1637] *La Géométrie.* English translation: *The Geometry.* Dover, New York 1954.

A.W. Dorodnov
[1947] On circular lunes quadrable with the use of ruler and compass (Russian). *Doklady Akad. Nauk. SSSR*, **58**, 965-968.

D.S. Dummit & R.M. Foote
[1991] *Abstract Algebra.* Prentice-Hall, Englewood Cliffs, New Jersey.

W. Dyck
[1883] Gruppentheoretische Studien II. *Math. Ann.* **22**, 70-108.

H.M. Edwards
[1984] *Galois Theory.* Springer-Verlag, New York.

G. Eisenstein
[1850] Über die Irreductibilität und einige Eigenschaften der Gleichung, von welcher die Theilung der ganzen Lemniscate abhängt. *J. Reine Angew. Math.* **39**, 224-287. *Werke* II, 536-555.

L. Euler
[1738] Observationes de theoremate quodam Fermatiano aliisque ad numeros primos spectantibus. *Novi Comm. Acad. Sci. Petrop.* **6**, 103-107, *Opera Omnia*, ser. 1, II, 1-5.

L. Euler
[1761] Theoremata arithmetica nova methodo demonstrata. *Novi Comm. Acad. Sci. Petrop.* **8**, 74-104, *Opera Omnia*, ser. 1, II, 531-555.

L. Euler
[1770] *Vollständige Einleitung zur Algebra.* English translation: *Elements of Algebra.* Longman, Orme and Co., London, 1849. Springer-Verlag, New York, 1984.

L. Euler

[1776] Formulae generales pro translatione quacumque corporum rigidorum. *Novi Comm. Acad. Sci. Petrop.* **20**, 189-207, *Opera omnia* ser. 2, IX, 84-98.

W. Feit and J.G. Thompson

[1963] The solvability of groups of odd order. *Pacific J. Math.* **13**, 1-255.

P. Fermat

[1629] Ad locos planos et solides isagoge. *Œuvres* 1, 92-103. English translation in D.E. Smith, *A Source Book in Mathematics* vol. 2, 389-396. Dover, New York, 1959.

P. Fermat

[1640] Letter to Frenicle, August(?) 1640. *Œuvres* 2, 205-206.

P. Fermat

[1640′] Letter to Frenicle, 18 October 1640. *Œuvres* 2, 209.

L. Fibonacci

[1225] *Flos Leonardo Bifolli Pisani super solutionibus quarundam quaestionum ad numerum et ad geometriam pertinentium. Scritti di Leonardo Pisano.* Baldassarre Boncompagni, Rome 1857-62.

E. Galois

[1831] Mémoire sur les conditions de résolubilité des équations par radicaux. *Écrits Mém. Math.* 43-71. English translation in Edwards [1984], 101-113.

C.F. Gauss

[1799] Demonstratio nova theorematis omnem functionem algebraicam rationalem integram unius variabilis in factores reales primi vel secundi gradus resolvi posse. Helmstedt dissertation. *Werke* 3, 1-30.

C.F. Gauss

[1801] *Disquisitiones Arithmeticae.* English translation (also entitled *Disquisitiones Arithmeticae*), Yale University Press, New Haven, 1966. Springer-Verlag, New York, 1986.

C.F. Gauss

[1816] Demonstratio nova altera theorematis omnem functionem algebraicam rationalem integram unius variabilis in factores reales primi vel secundi gradus resolvi posse. *Comm. Recentiores (Gottingae)* **3**, 107-142. *Werke* 3, 31-56.

C. Goldbach

[1730] Letter to Euler, 20/31 July. *Correspondance Mathématique et Physique* (Ed. P.-H. Fuss). Johnson Reprint Corporation, New York, 1968.

C.R. Hadlock

[1978] *Field Theory and Its Classical Problems.* Math. Assoc. Amer., Washington.

W.R. Hamilton
[1837] Theory of conjugate functions, or algebraic couples, with a preliminary and elementary essay on algebra as the science of pure time. *Trans. Roy. Irish Acad.*, XVII, part II, 293-422.

T.L. Heath
[1912] *The Works of Archimedes.* Cambridge U. Press. Reprinted by Dover, New York.

T.L. Heath
[1925] *The Thirteen Books of Euclid's Elements.* Cambridge U. Press. Reprinted by Dover, New York, 1956.

C. Hermite
[1873] Sur la fonction exponentielle. *Comp. Rend.* **77**, 18-24, 74-79, 226-233, 285-293. *Œuvres* 3, 150-181.

K. Ireland & M. Rosen
[1982] *A Classical Introduction to Modern Number Theory.* Springer-Verlag, New York.

C. Jordan
[1870] *Traité des substitutions et des équations algébriques.* Gauthier-Villars, Paris.

B.M. Kiernan
[1971] *The Development of Galois Theory from Lagrange to Artin.* Arch. Hist. Exact Sci. **3**, 40-154.

F. Klein
[1872] *Vergleichende Betrachtungen über neuere geometrische Forschungen (Erlanger Programm).* Akademische Verlagsgesellschaft, Leipzig 1974, *Ges. Math. Abhandlungen* 1, 460-497.

F. Klein
[1876] Über binäre Formen mit linearen Transformationen in sich selbst. *Math. Ann.* **9**, 183-208.

F. Klein
[1884] *Vorlesungen über das Ikosaeder und die Auflösung der Gleichungen vom fünften Grade.* Teubner, Leipzig. English translation: *Lectures on the Icosahedron*, Dover, New York, 1956.

H. Koch
[1991] *Introduction to Classical Mathematics* I. Kluwer, Dordrecht.

L. Kronecker
[1853] Über die algebraisch auflösbaren Gleichungen. *Ber. König. Akad. Wiss. Berlin*, 365-374, *Werke* IV, 1-11.

L. Kronecker
[1854] Mémoire sur les facteurs irréductibles de l'expression $x^n - 1$. *J. de Math. Pures et Appl.* **19**, 177-192, *Werke* I, 75-97.

L. Kronecker

[1870] Auseinandersetzung einiger Eigenschaften der Klassenzahl idealer complexer Zahlen. *Monatsber. Königl. Akad. Wiss. Berlin*, 881-889. *Werke* I, 271-282.

L. Kronecker

[1887] Ein Fundamentalsatz der allgemeinen Arithmetik. *J. Reine Angew. Math.* **100**, 490-510. *Werke* III, 209-240.

L. Kronecker

[1901] *Vorlesungen über Zahlentheorie* I. Springer-Verlag, Berlin, 1978.

J.L. Lagrange

[1771] Réflexions sur la résolution algébrique des équations. *Nouv. Mém. l'Acad. Berlin. Œuvres* 3, 205-421.

K. Lamotke

[1986] *Regular Solids and Isolated Singularities.* Vieweg, Braunschweig, 1986.

P.S. Laplace

[1795] Leçons de Mathematiques données à l'École Normale, en 1795. *J.École Polytechn.* II (Paris 1812), 1-278.

A.M. Legendre

[1798] *Essai sur la théorie des nombres.* Paris.

J. Libbrecht

[1973] *Chinese Mathematics in the Thirteenth Century.* MIT Press, Cambridge, Mass.

F. Lindemann

[1882] Über die Zahl π. *Math. Ann.* **20**, 213-225.

J. Liouville

[1844] Remarques relatives à des classes très-étendues de quantités dont la valeur n'est ni rationelle ni même réducible à des irrationelles algébriques. *Comp. rend.* **18**, 883-885. Nouvelle démonstration d'un théorème sur les irrationelles algébriques. *Ibid.*, 910-911.

I. Newton

[1707] *Arithmetica universalis.* Cambridge. English translation in *The Mathematical Works of Isaac Newton*, vol. II (Ed. D.T. Whiteside), Johnson Reprint Co., 1964.

E. Noether

[1927] Abstrakter Aufbau der Idealtheorie in algebraischen Zahl- und Funktionenkörpern. *Math. Ann.* **96**, 26-61.

E. Noether

[1929] Hyperkomplexe Grössen und Darstellungstheorie. *Math. Zeit.* **30**, 641-692.

L. Pacioli
[1509] *De Divina Proportione.* A. Paganius Paganinus, Venice. Reprinted by Mediobanca di Milano, Milan, 1956

B. Pascal
[1654] Traité du triangle arithmétique, avec quelques autres petits traités sur la même manière. English translation in *Great Books of the Western World: Pascal.* Encyclopedia Brittanica, London 1952, 447-473.

J. Rotman
[1990] *Galois Theory.* Springer-Verlag, New York.

P. Ruffini
[1799] *Teoria Generale delle Equazioni. Opere* I, 1-334.

P. Ruffini
[1813] Riflessioni intorno alla soluzione delle equazioni algebraiche generali. *Opere* II, 155-268.

J.-P. Serre
[1992] *Topics in Galois Theory.* Jones and Bartlett, Boston.

L. Shafarevich
[1954] Construction of fields of algebraic numbers with given solvable Galois group. (Russian) *Izv. Akad. Nauk SSSR* **18**, 525- 578. English translation in *Amer. Math. Soc. Transl.* **4**, (1956), 185-237.

M. Sono
[1917] On Congruences I, II, III, IV. *Mem. College of Sci., Kyoto Imperial U.* **2**, **3**.

E. Steinitz
[1910] *Algebraischer Theorie der Körper. J. Reine Angew. Math.* Reprinted by Chelsea, New York, 1950.

S. Stevin
[1585] *L'arithmétique.* Leyden. Abridgement in *Principal Works of Simon Stevin,* vol. IIB, 477-708.

J.C. Stillwell
[1989] *Mathematics and Its History.* Springer-Verlag, New York.

J-P. Tignol
[1988] *Galois' Theory of Algebraic Equations.* Longman-Wiley, New York.

B.L. van der Waerden
[1931] *Moderne Algebra.* Springer-Verlag, Berlin. English translation: *Modern Algebra.* Ungar, New York, 1949.

B.L. van der Waerden
[1975] On the sources of my book *Moderne Algebra. Hist. Math.* **2**, 31-40.

F. Viète
[1591] De aequationum recognitione et emendatione. *Opera*, 82-162. English translation in *The Analytic Art*, Kent State U. Press, Kent, Ohio, 1983.

M.J. Wallenius
[1766] Åbo dissertation, cited in Heath's *A History of Greek Mathematics*, vol. 1, Dover 1981, 200.

P.L. Wantzel
[1837] Recherches sur les moyens de reconnaitre si un problème peut se resoudre avec la règle et le compass. *J. Math.* **2**, 366-372.

H. Weber
[1886] Theorie der Abelschen Zahlkörper. *Acta. Math.* **8**, 193-263.

H. Weber
[1893] Leopold Kronecker. *Math. Ann.* **43**, 1-25.

A. Weil
[1984] *Number Theory. An Approach Through History.* Birkhäuser, Basel.

K. Weierstrass
[1874] *Einleitung in die Theorie der analytischen Funktionen.* Summer Semester 1874. Notes by G. Hettner. Mathematische Institut der Universität Göttingen.

H. Weyl
[1935] Emmy Noether. *Scripta Math.* III, 3, 201-220.

Index

(continued from page ii)

Halmos: Naive Set Theory.
Hämmerlin/Hoffmann: Numerical Mathematics. *Readings in Mathematics.*
Harris/Hirst/Mossinghoff: Combinatorics and Graph Theory.
Hartshorne: Geometry: Euclid and Beyond.
Hijab: Introduction to Calculus and Classical Analysis.
Hilton/Holton/Pedersen: Mathematical Reflections: In a Room with Many Mirrors.
Hilton/Holton/Pedersen: Mathematical Vistas: From a Room with Many Windows.
Iooss/Joseph: Elementary Stability and Bifurcation Theory. Second edition.
Isaac: The Pleasures of Probability. *Readings in Mathematics.*
James: Topological and Uniform Spaces.
Jänich: Linear Algebra.
Jänich: Topology.
Jänich: Vector Analysis.
Kemeny/Snell: Finite Markov Chains.
Kinsey: Topology of Surfaces.
Klambauer: Aspects of Calculus.
Lang: A First Course in Calculus. Fifth edition.
Lang: Calculus of Several Variables. Third edition.
Lang: Introduction to Linear Algebra. Second edition.
Lang: Linear Algebra. Third edition.
Lang: Undergraduate Algebra. Second edition.
Lang: Undergraduate Analysis.
Lax/Burstein/Lax: Calculus with Applications and Computing. Volume 1.
LeCuyer: College Mathematics with APL.
Lidl/Pilz: Applied Abstract Algebra. Second edition.
Logan: Applied Partial Differential Equations.
Macki-Strauss: Introduction to Optimal Control Theory.
Malitz: Introduction to Mathematical Logic.
Marsden/Weinstein: Calculus I, II, III. Second edition.
Martin: Counting: The Art of Enumerative Combinatorics.
Martin: The Foundations of Geometry and the Non-Euclidean Plane.
Martin: Geometric Constructions.
Martin: Transformation Geometry: An Introduction to Symmetry.
Millman/Parker: Geometry: A Metric Approach with Models. Second edition.
Moschovakis: Notes on Set Theory.
Owen: A First Course in the Mathematical Foundations of Thermodynamics.
Palka: An Introduction to Complex Function Theory.
Pedrick: A First Course in Analysis.
Peressini/Sullivan/Uhl: The Mathematics of Nonlinear Programming.
Prenowitz/Jantosciak: Join Geometries.
Priestley: Calculus: A Liberal Art. Second edition.
Protter/Morrey: A First Course in Real Analysis. Second edition.
Protter/Morrey: Intermediate Calculus. Second edition.
Roman: An Introduction to Coding and Information Theory.
Ross: Elementary Analysis: The Theory of Calculus.
Samuel: Projective Geometry. *Readings in Mathematics.*
Scharlau/Opolka: From Fermat to Minkowski.
Schiff: The Laplace Transform: Theory and Applications.